T0332877

# Mathematics of Harmony as a New Interdisciplinary Direction and "Golden" Paradigm of Modern Science

## Volume 2

Algorithmic Measurement Theory, Fibonacci and Golden Arithmetic's and Ternary Mirror-Symmetrical Arithmetic

K&E Series on Knots and Everything — Vol. 68

# Mathematics of Harmony as a New Interdisciplinary Direction and "Golden" Paradigm of Modern Science

### Volume 2
#### Algorithmic Measurement Theory, Fibonacci and Golden Arithmetic's and Ternary Mirror-Symmetrical Arithmetic

## Alexey Stakhov

International Club of the Golden Section, Canada & Academy of Trinitarism, Russia

**World Scientific**

EW JERSEY · LONDON · SINGAPORE · BEIJING · SHANGHAI · HONG KONG · TAIPEI · CHENNAI · TOKYO

*Published by*

World Scientific Publishing Co. Pte. Ltd.

5 Toh Tuck Link, Singapore 596224

*USA office:* 27 Warren Street, Suite 401-402, Hackensack, NJ 07601

*UK office:* 57 Shelton Street, Covent Garden, London WC2H 9HE

**Library of Congress Cataloging-in-Publication Data**
Names: Stakhov, A. P. (Alekseĭ Petrovich), author.
Title: Mathematics of harmony as a new interdisciplinary direction and
    "golden" paradigm of modern science / Alexey Stakhov.
Description: Hackensack, New Jersey : World Scientific, [2020] | Series: Series on knots and
    everything, 0219-9769 ; vol. 68 | Includes bibliographical references. | Contents:
    Volume 1. The golden section, Fibonacci numbers, Pascal triangle, and Platonic solids --
    Volume 2. Algorithmic measurement theory, Fibonacci and golden arithmetic and
    ternary mirror-symmetrical arithmetic -- Volume 3. The "golden" paradigm of modern science.
Identifiers: LCCN 2020010957 | ISBN 9789811207105 (v. 1 ; hardcover) |
    ISBN 9789811213465 (v. 2 ; hardcover) | ISBN 9789811206375 (v. 1 ; ebook) |
    ISBN 9789811213496 (v. 3 ; hardcover) | ISBN 9789811206382 (v. 1 ; ebook other) |
    ISBN 9789811213472 (v. 2 ; ebook) | ISBN 9789811213489 (v. 2 ; ebook other) |
    ISBN 9789811213502 (v. 3 ; ebook) | ISBN 9789811213519 (v. 3 ; ebook other)
Subjects: LCSH: Fibonacci numbers. | Golden section. | Mathematics--History. | Science--Mathematics.
Classification: LCC QA246.5 .S732 2020 | DDC 512.7/2--dc23
LC record available at https://lccn.loc.gov/2020010957

**British Library Cataloguing-in-Publication Data**
A catalogue record for this book is available from the British Library.

For any available supplementary material, please visit
https://www.worldscientific.com/worldscibooks/10.1142/11644#t=suppl

Desk Editor: Liu Yumeng

Typeset by Stallion Press
Email: enquiries@stallionpress.com

Printed in Singapore

*In fond memory of Yuri Alekseevich Mitropolskiy*
*and*
*Alexander Andreevich Volkov*

# Contents

# Preface to the Three-Volume Book

## Continuity in the Development of Science

Scientific and technological progress has a long history and passed in its historical development several stages: The Babylonian and Ancient Egyptian culture, the culture of Ancient China and Ancient India, the Ancient Greek culture, the Middle Ages, the Renaissance, the Industrial Revolution of the 18th century, the Great Scientific Discoveries of the 19th century, the Scientific and Technological Revolution of the 20th century and finally the 21st century, which opens a new era in the history of mankind, the *Era of Harmony*.

Although each of the mentioned stages has its own specifics, at the same time, every stage necessarily includes the content of the preceding stages. This is called the *continuity* in the development of science.

It was during the ancient period, a number of the fundamental discoveries in mathematics were made. They exerted a determining influence on the development of the material and spiritual culture. We do not always realize their importance in the development of mathematics, science, and education. To the category of such discoveries, first of all, we must attribute the *Babylonian numeral system with the base* 60 and the *Babylonian positional principle of number representation*, which is the foundation of the, *decimal, binary, ternary,* and other positional numeral systems. We must add to this list the *trigonometry* and the *Euclidean geometry,* the *incommensurable segments* and the *theory of irrationality,* the *golden section* and *Platonic solids*, the *elementary number theory* and the *mathematical theory of measurement*, and so on.

The *continuity* can be realized in various forms. One of the essential forms of its expression are the fundamental scientific ideas, which permeate all stages of the scientific and technological progress and influence various areas of science, art, philosophy, and technology. The idea of *Harmony*, connected with the *golden section*, belongs to the category of such fundamental ideas.

According to B.G. Kuznetsov, the researcher of Albert Einstein's creativity, the great physicist piously believed that science, physics in particular, always had its eternal fundamental goal *"to find in the labyrinth of the observed facts the objective harmony"*. The deep faith of the outstanding physicist in the existence of the universal laws of the *Harmony* is evidenced by another well-known Einstein's statement: *"The religiousness of the scientist consists in the enthusiastic admiration for the laws of the Harmony"* (the quote is taken from the book *Meta-language of Living Nature* [1], written by the outstanding Russian architect Joseph Shevelev, known for his research in the field of *Harmony* and the *golden section* [1–3]).

## Pythagoreanism and Pythagorean MATHEM's

By studying the sources of the origin of mathematics, we inevitably come to Pythagoras and his doctrine, named the *Pythagoreanism* (see Wikipedia article *Pythagoreanism*, the Free Encyclopedia). As mentioned in Wikipedia, the *Pythagoreanism* originated in the 6th century BC and was based on teachings and beliefs of Pythagoras and his followers called the Pythagoreans. Pythagoras established the first Pythagorean community in Croton, Italy. The Early Pythagoreans espoused a rigorous life and strict rules on diet, clothing and behavior.

According to tradition, *Pythagoreans* were divided into two separate schools of thought: the *mathematikoi* (*mathematicians*) and the *akousmatikoi* (*listeners*). The *listeners* had developed the religious and ritual aspects of *Pythagoreanism*; the *mathematicians* studied the four Pythagorean MATHEMs: *arithmetic*, *geometry, spherics,* and *harmonics.* These MATHEMs, according to Pythagoras, were the main composite parts of mathematics. Unfortunately, the Pythagorean MATHEM of the *harmonics* was lost in mathematics during the process of its historical development.

## Proclus Hypothesis

The Greek philosopher and mathematician Proclus Diadoch (412–485 AD) put forth the following unusual hypothesis concerning Euclid's *Elements.* Among Proclus's mathematical works, his *Commentary on the Book I of Euclid's Elements* was the most well known. In the commentary, he puts forth the following unusual hypothesis.

It is well known that Euclid's *Elements* consists of 13 books. In those, XIII$^{\text{th}}$ book, that is, the concluding book of the *Elements,* was devoted to the description of the geometric theory of the five *regular polyhedra,* which had played a dominant role in *Plato's cosmology* and is known in modern science under the name of the *Platonic solids.*

Proclus drew special attention to the fact that the concluding book of the *Elements* had been devoted to the *Platonic solids.* Usually, the most important material, of the scientific work is placed in its final part. Therefore, by placing *Platonic solids* in Book XIII, that is, in the concluding book of his *Elements,* Euclid clearly pointed out on main purpose of writing his *Elements.* As the prominent Belarusian philosopher Edward Soroko points out in [4], according to Proclus, Euclid *"had created his Elements allegedly not for the purpose of describing geometry as such, but with purpose to give the complete systematized theory of constructing the five Platonic solids; in the same time Euclid described here some latest achievements of mathematics".*

It is for the solution of this problem (first of all, for the creation of geometric theory of *dodecahedron*), Euclid already in Book II introduces Proposition II.11, where he describes the *task of dividing the segment in the extreme and mean ratio* (the *golden section*), which then occurs in other books of the *Elements,* in particular in the concluding book (XIII Book).

But the *Platonic solids* in *Plato's cosmology* expressed the *Universal Harmony* which was the main goal of the ancient Greeks science. With such consideration of the *Proclus hypothesis*, we come to the surprising conclusion, which is unexpected for many historians of mathematics. According to the *Proclus hypothesis*, it turns out that from Euclid's *Elements*, two branches of mathematical sciences had originated: the **Classical Mathematics**, which included the *Elements* of the *axiomatic approach* (Euclidean axioms), *the elementary number theory*, and *the theory of irrationalities*, and the Mathematics of Harmony, which was based on the geometric *"task of dividing the segment in the extreme and mean ratio"* (the *golden section*) and also on the theory of the *Platonic solids*, described by Euclid in the concluding Book XIII of his *Elements*.

### The Statements by Alexey Losev and Johannes Kepler

What was the main idea behind ancient Greek science? Most researchers give the following answer to this question: **The idea of Harmony connected to the *golden section***. As it is known, in ancient Greek philosophy, *Harmony* was in opposition to the *Chaos* and meant the organization of the Universe, the Cosmos. The outstanding Russian philosopher Alexey Losev (1893–1988), the researcher in the aesthetics of the antiquity and the Renaissance, assesses the main achievements of the ancient Greeks in this field as follows [5]:

> *"From Plato's point of view, and in general in the terms of the entire ancient cosmology, the Universe was determined as the certain proportional whole, which obeys to the law of the harmonic division, the golden section ... The ancient Greek system of the cosmic proportion in the literature is often interpreted as the curious result of the unrestrained and wild imagination. In such explanation we see the scientific helplessness of those, who claim this. However, we can understand this historical and aesthetic phenomenon only in the connection with the holistic understanding of history, that is, by using the dialectical view on the culture and by searching for the answer in the peculiarities of the ancient social life."*

Here, Losev formulates the *"golden" paradigm* of ancient cosmology. This paradigm was based upon the fundamental ideas of ancient science that are sometimes treated in modern science as the "*curious result of the unrestrained and wild imagination*". First of all, we are talking about the *Pythagorean Doctrine of the Numerical Universal Harmony* and *Plato's Cosmology* based on the *Platonic solids*. By referring to the geometrical structure of the Cosmos and its mathematical relations, which express the Cosmic Harmony, the Pythagoreans had anticipated the modern mathematical basis of the natural sciences, which began to develop rapidly in the 20th century. Pythagoras's and Plato's ideas about the Cosmic Harmony proved to be immortal.

Thus, the idea of Harmony, which underlies the ancient Greek doctrine of Nature, was the main "paradigm" of the Greek science, starting from Pythagoras and ending with Euclid. This paradigm relates directly to the *golden section* and the *Platonic solids*, which are the most important Greek geometric discoveries for the expression of the Universal Harmony.

Johannes Kepler (1571–1630), the prominent astronomer and the author of "Kepler's laws", expressed his admiration with the *golden ratio* in the following words [6]:

> "*Geometry has the two great treasures: the first of them is the theorem of Pythagoras; the second one is the division of the line in the extreme and mean ratio. The first one we may compare to the measure of the gold; the second one we may name the precious stone.*"

We should recall again that the ancient *task of dividing line segment in extreme and mean ratio* is Euclidean language for the *golden section*!

The enormous interest in this problem in modern science is confirmed by the rather impressive and far from the complete list of books and articles on this subject, published in the second half of the 20th century and the beginning of the 21st century [1–100].

### Ancient Greeks Mathematical Doctrine of Nature

According to the outstanding American historian of mathematics, Morris Kline [101], the main contribution of the ancient Greeks is the one "*which had the decisive influence on the entire subsequent culture, was that they took up the study of the laws of Nature"*. The main conclusion, from Morris Kline's book [101] is the fact that the ancient Greeks proposed the innovative concept of the Cosmos, in which everything was subordinated to the mathematical laws. Then the following question arises: during which time this concept was developed? The answer to this question is also addressed in Ref. [101].

According to Kline [101], the innovative concept of the Cosmos based on the mathematical laws, was developed by the ancient Greeks in the period from VI to III centuries BC. But according to the prominent Russian mathematician academician A.N. Kolmogorov [102], in the same period in ancient Greece, *"the mathematics was created as the independent science with the clear understanding of the uniqueness of its method and with the need for the systematic presentation of its basic concepts and proposals in the fairly general form."* But then, the following important question, concerning the history of the original mathematics arises: was there any relationship between the process of creating the mathematical theory of Nature, which was considered as the goal and the main achievement of ancient Greek science [101], and the process of creating mathematics, which happened in ancient Greece in the same period [102]? It turns out that such connection, of course, existed. Furthermore, it can be argued that these processes actually coincided, that is, the processes of the creation of mathematics by the ancient Greeks [102], and their doctrine of Nature, based on the mathematical principles [101], were one and the same processes. And the most vivid embodiment of the process of the *Mathematization of Harmony* [68] happened in Euclid's *Elements*, which was written in the third century BC.

## Introduction of the Term *Mathematics of Harmony*

In the late 20th century, to denote the mathematical doctrine of Nature, created by the ancient Greeks, the term *Mathematics of Harmony* was introduced. It should be noted that this term was chosen very successfully because it reflected the main idea of the ancient Greek science, the *Harmonization of Mathematics* [68]. For the first time, this term was introduced in the small article "Harmony of spheres", placed in *The Oxford Dictionary of Philosophy* [103]. In this article, the concept of *Mathematics of Harmony* was associated with the *Harmony of spheres*, which was, also called in Latin as "*harmonica mundi*" or "*musica mundana*" [10]. The *Harmony of spheres* is the ancient and medieval doctrine on the musical and mathematical structure of the Cosmos, which goes back to the Pythagorean and Platonic philosophical traditions.

Another mention about the *Mathematics of Harmony* in the connection to the ancient Greek mathematics is found in the book by Vladimir Dimitrov, *A New Kind of Social Science*, published in 2005 [44]. It is important to emphasize that in Ref. [44], the concept of *Mathematics of Harmony* is directly associated with the *golden section*, the most important mathematical discovery of the ancient science in the field of Harmony. This discovery at that time was called "*dividing a segment into the extreme and mean ratio*" [32].

From Refs. [44, 45], it is evident that prominent thinkers, scientists and mathematicians took part in the development of the *Mathematics of Harmony* for several millennia: Pythagoras, Plato, Euclid, Fibonacci, Pacioli, Kepler, Cassini, Binet, Lucas, Klein, and in the 20th century the well-known mathematicians Coxeter [7], Vorobyov [8], Hoggatt [9], Vaida [11], Knuth [123], and so on. And we cannot ignore this historical fact.

## Fibonacci Numbers

The Fibonacci numbers, introduced into Western European mathematics in the 13th century by the Italian mathematician Leonardo of Pisa (known by the nickname Fibonacci), are closely related to the *golden ratio*. Fibonacci numbers from the numerical sequence, which starts with two units, and then each subsequent Fibonacci number is the sum of the two previous ones: $1, 1, 2, 3, 5, 8, 13, 21, 34, 55, \ldots$. The ratio of the two neighboring Fibonacci numbers in the limit tends to be the *golden ratio*.

The mathematical theory of Fibonacci numbers has been further developed in the works of the French mathematicians of the 19th century Binet (*Binet formula*) and Lucas (*Lucas numbers*). As mentioned above, in the second half of the 20th century, this theory was developed in the works of the Canadian geometer, Donald Coxeter [7], the Soviet mathematician, Nikolay Vorobyov [8], the American mathematician, Verner Hoggatt [9] and the English mathematician, Stefan Vajda [11], the outstanding American mathematician, Donald Knuth [123], and so on.

The development of this direction ultimately led to the emergence of the *Mathematics of Harmony* [6], a new interdisciplinary direction of modern science that relates to modern mathematics, computer science, economics, as well as to all theoretical natural sciences. The works of the well-known mathematicians, Coxeter [7], Vorobyov [8], Hoggatt [9], Vaida [11], Knuth [123], and others, as well as the study of Fibonacci mathematicians, members of the American Fibonacci Association, became the beginning of the process of *Harmonization of Mathematics* [68], which continues actively in the 21st century. And this process is confirmed by a huge number of books and articles in the field of the *golden section* and *Fibonacci numbers* published in the second half of the 20th century and the beginning of the 21st century [1–100].

## Sources of the Present Three-Volume Book

The differentiation of modern science and its division into separate spheres do not allow us often to see the general picture of science and the main trends in its development. However, in science, there exist research objects that combine disparate scientific facts into a single whole. *Platonic solids* and the *golden section* are attributed to the category of such objects. The ancient Greeks elevated them to the level of *"the main harmonic figures of the Universe"*. For centuries or even millennia, starting from Pythagoras, Plato and Euclid, these geometric objects were the object of admiration and worship of the outstanding minds of mankind, during Renaissance, Leonardo da Vinci, Luca Pacoli, Johannes Kepler, in the 19th century, Zeising, Lucas, Binet and Klein. In the 20th century, the interest in these mathematical objects increased significantly, thanks to the research of the Canadian geometer, Harold Coxeter [7], the Soviet mathematician Nikolay Vorobyov [8] and the American mathematician Verner Hoggatt [9], whose works in the field of the Fibonacci numbers began the process of the "Harmonization of Mathematics". The development of this direction led to the creation of the *Mathematics of Harmony* [6] as a new interdisciplinary trend of modern science.

The newest discoveries in the various fields of modern science, based on the *Platonic solids*, the *golden section* and the *Fibonacci numbers*, and new scientific discoveries and mathematical results, related to the *Mathematics of Harmony* (*quasicrystals* [115], *fullerenes* [116], the new geometric theory of phyllotaxis (*Bodnar's geometry*) [28], the *general theory of the hyperbolic functions* [75, 82], the *algorithmic measurement theory* [16], the *Fibonacci and golden ratio codes* [6], the *"golden" number theory* [94], the *"golden" interpretation of the special theory of relativity* and the *evolution of the Universe* [87], and so on) create an overall picture of the movement of modern science towards the *"golden" scientific revolution*, which is one of the characteristic trends in the development of modern science. The sensational information about the experimental discovery of the golden section in the quantum world as a result of many years of research, carried out at the Helmholtz–Zentrum Berlin für Materialien und Energie (HZB) (Germany), the Oxford and Bristol Universities and the Rutherford Appleton Laboratory (UK), is yet another confirmation of the movement of the theoretical physics to the *golden section* and the *Mathematics of Harmony* [6].

For the first time, this direction was described in the book by Stakhov A.P., assisted by Scott Olsen, *The Mathematics of Harmony. From Euclid to Contemporary Mathematics and Computer Science*, World Scientific, 2009 [6].

In 2006, the Russian Publishing House, "Piter" (St. Petersburg) published the book, *Da Vinci Code and Fibonacci numbers* [46] (Alexey Stakhov, Anna Sluchenkova and Igor Shcherbakov were the authors of the book). This book was one of the first Russian books in this field. Some aspects of this direction are reflected in the following authors' books, published by Lambert Academic Publishing (Germany) and World Scientific (Singapore):

- Alexey Stakhov, Samuil Aranson, *The Mathematics of Harmony and Hilbert's Fourth Problem. The Way to Harmonic Hyperbolic and Spherical Worlds of Nature.* Germany: Lambert Academic Publishing, 2014 [51].
- Alexey Stakhov, Samuil Aranson, Assisted by Scott Olsen, *The "Golden" Non-Euclidean Geometry,* World Scientific, 2016 [52].
- Alexey Stakhov, *Numeral Systems with Irrational Bases for Mission-Critical Applications,* World Scientific, 2017 [53].

These books are fundamental in the sense that they are the first books in modern science, devoted to the description of the theoretical foundations and applications of the following new trends in modern science: the *history of the golden section* [78], the *Mathematics of Harmony* [6], the *"Golden" Non-Euclidean geometry* [52], ascending to Euclid's *Elements,* and also the *Numeral Systems with Irrational bases,* ascending to the Babylonian positional numeral system, the decimal and binary system and Bergman's system [54].

These books discuss the problems, which in modern mathematics are considered long resolved and therefore are not included in the circle of the studies of mathematicians, namely the new mathematical theory of measurement called the *Algorithmic Measurement Theory* [16, 17], the *Mathematics of Harmony* [6] as a new kind of elementary mathematics that has a direct relationship to the foundations of the mathematics and mathematical education, the new class of the elementary functions called the *hyperbolic Fibonacci and Lucas functions* [64, 75, 82] and finally, the new ways of real numbers representation, and the new binary and ternary arithmetic's [55, 72], which have the fundamental interest for computer science and digital metrology.

In 2010, the Odesa I.I. Mechnikov National University (Ukraine) hosted the *International Congress on the Mathematics of Harmony.* The main goal of the Congress was to consolidate the priority of Slavic science in the development of this important trend and acquaint the scientific community with the main trends of the development of the Mathematics of Harmony as the new interdisciplinary direction of modern science.

In the recent years, the new unique books on the problems of Harmony and the history of the golden section have been published:

- *The Prince of Wales. Harmony. A New Way of Looking at our World* (*coauthors Tony Juniper and Ian Skelly*). An Imprint of HarperCollins Publisher, 2010 [49].
- Hrant Arakelian, *Mathematics and History of the Golden Section.* Moscow, Publishing House "Logos", 2014 [50].

For the last 30 years, *Charles, The Prince of Wales*, had been known around the world as one of the most forceful advocates for the environment. During that period, he focused on many different aspects of our lives, when we continually confront with the real life from new angles of view and search original approaches. Finally, in *Harmony* (2010) [49], The Prince of Wales and his coauthors laid out their thoughts on the planet, by offering an in-depth look into its future. Here, we see a dramatic call to the action and an inspirational guide on the relationship of mankind with Nature throughout history. The Prince of Wales's *Harmony* (2010) [49] is an illuminating look on how we must reconnect with our past in order to take control of our future.

The 2014 book [50] by the Armenian philosopher and physicist *Hrant Arakelian* is devoted to the *golden section* and to the complexity of problems connected with it. The book consists of two parts. The first part is devoted to the mathematics of the golden section and the second part to the history of the golden section. Undoubtedly, Arakelian's 2014 book is one of the best modern books devoted to mathematics and the history of the golden section.

The *International Congress on Mathematics of Harmony* (Odessa, 2010) and the above-mentioned books by *The Prince of Wales* and Armenian philosopher *Hrant Arakelian* are brilliant confirmation of the fact that in modern science, the interest in the mathematics of the golden section and its history increases and further development of the Mathematics of Harmony can lead to revolutionary transformations in modern mathematics and science on the whole.

Why did the author decide to write the three-volume book *The Mathematics of Harmony as a New Interdisciplinary Direction and "Golden" Paradigm of Modern Science*? It should be noted that the author and other famous authors in this field published many original books and articles in this scientific direction. However, all the new results and ideas, described in the above-mentioned publications of Alexey Stakhov, Samuil Aranson, Charles, The Prince of Wales, Hrant Arakelian and other authors are scattered in their numerous articles and books, which makes it difficult to understand their fundamental role in the development of the modern mathematics, computer science and theoretical natural sciences on the whole.

This role is most clearly reflected in the following citations taken from *Harmony* by the Prince of Wales (2010) [49]:

*"This is a call to revolution. The Earth is under threat. It cannot cope with all that we demand of it. It is losing its balance and we humans are causing this to happen."*

The following quote, placed on the back cover of Prince of Wales's *Harmony* [49], develops this thought:

*"We stand at an historical moment; we face a future where there is a real prospect that if we fail the Earth, we fail humanity. To avoid such an outcome, which will comprehensively destroy our children's future or even our own, we must make choices now that carry monumental implications."*

Thus, *The Prince of Wales* has considered his 2010 book, *Harmony. A New Way of Looking at our World*, as a call to the revolution in modern science, culture and education. The same point of view is expressed in the above-mentioned books by Stakhov and Aranson [6, 46, 51–53]. Comparing the books of *Prince of Wales* [49] and *Hrant Arakelian* [50] to the 2009, 2016 and 2017 books of Alexey Stakhov and Samuil Aranson [6, 51–53], one can only be surprised how deeply all these books, written in different countries and continents, coincide in their ideas and goals.

Such an amazing coincidence can only be explained by the fact that in modern science, there is an urgent need to return to the "harmonious ideas" of Pythagoras, Plato and Euclid that permeated across the ancient Greek science and culture. The *Harmony* idea, formulated in the works of the Greek scholars and reflected in Euclid's *Elements* turned out to be immortal!

We can safely say that the above-mentioned books by Stakhov and Aranson (2009, 2016, 2017) [6, 51–53], the book by The Prince of Wales with the coauthors (2010) [49] and book by Arakelian (2014) [50] are the beginning of a revolution in modern science. The essence of this revolution consists, in turning to the fundamental ancient Greek idea of the *Universal Harmony*, which can save our Earth and humanity from the approaching threat of the destruction of all mankind.

It was this circumstance that led the author to the idea of writing the three-volume book *Mathematics of Harmony as a New Interdisciplinary Direction and "Golden" Paradigm of Modern Science*, in which the most significant and fundamental scientific results and ideas, formulated by the author and other authors (The Prince of Wales, Hrant Arakelian, Samuil Aranson and others) in the process of the development of this scientific direction, will be presented in a popular form, accessible to students of universities and colleges and teachers of mathematics, computer science, theoretical physics and other scientific disciplines.

## Structure and the Main Goal of the Three-Volume Book

The book consists of three volumes:

- *Volume I. The Golden Section, Fibonacci Numbers, Pascal Triangle and Platonic Solids.*
- *Volume II. Algorithmic Measurement Theory, Fibonacci and Golden Arithmetic and Ternary Mirror-Symmetrical Arithmetic.*
- *Volume III. The "Golden" Paradigm of Modern Science: Prerequisite for the "Golden" Revolution in the Mathematics, the Computer Science, and Theoretical Natural Sciences.*

Because the *Mathematics of Harmony* goes back to the "harmonic ideas" of Pythagoras, Plato and Euclid, the publication of such a three-volume book will promote the introduction of these "harmonic ideas" into modern education, which is important for more in-depth understanding of the ancient conception of the *Universal Harmony* (as the main conception of ancient Greek science) and its effective applications in modern mathematics, science and education.

The main goal of the book is to draw the attention of the broad scientific community and pedagogical circles to the *Mathematics of Harmony*, which is a new kind of elementary mathematics and goes back to Euclid's *Elements*. The book is of interest for the modern mathematical education and can be considered as the "golden" paradigm of modern science on the whole.

The book is written in a popular form and is intended for a wide range of readers, including schoolchildren, school teachers, students of colleges and universities and their teachers, and also scientists of various specializations, who are interested in the history of mathematics, Platonic solids, golden section, Fibonacci numbers and their applications in modern science.

# Introduction

Volume II is devoted to the discussion of two fundamental problems of science and mathematics, the *problem of measurement* and the *problem of numeral systems*, their relationship with the development of science and their historical role in the development, first of all, of contemporary mathematics and computer science by taking into consideration the contemporary achievements in mathematical theory of measurement and numeral systems.

As it is known, a set of rules, used by the ancient Egyptian land surveyors, was the first *measurement theory*. From this *measurement theory*, as the ancient Greeks testify, there originated the *geometry*, which takes its origin (and title) in the problem of *earth measuring*.

Already in the ancient Greece, the mathematical problems of *geometry* (that is, *earth measuring*) were the main focus of ancient mathematics. The science of measurement, related to *geometry*, was developing primarily as a *mathematical theory*. It is during this period that the discovery of the *incommensurable segments* and the formulation of *Eudoxus' exhaustion method*, to which the *number theory* as well as the *integral and differential calculus* go back in its origin, were made.

By basing on these important mathematical discoveries, which had the relation to *measurement*, the Bulgarian mathematician academician Ljubomir Iliev, the leader of the Bulgarian mathematical community, asserted that "*during the first epoch of its development, from antiquity to until the discovery of differential calculus, mathematics, by studying primarily problems of measurement, did created the Euclidean geometry and number theory"* [137].

In 1991, the Publishing House "Science" (the main Russian edition of the physical and mathematical literature) has published the book, *Mathematics in its Historical Development* [102], written by the outstanding Russian mathematician academician Andrey Kolmogorov (1903–1987). By discussing the period of the origin of mathematics, academician Kolmogorov pays attention to the following features of this period:

> *"The counting of objects at the earliest stages of development of culture led to the creation of the simplest concepts of arithmetic of natural numbers. Only on the basis of the developed system of oral numeration, the written numeral systems arise, and gradually methods of performing four arithmetic operations over natural numbers are developed ...*
>
> *The demand for measurement (the amount of grain, the length of the road, etc.) leads to the appearance of the names and symbols of the most widespread fractional numbers and the development of methods for performing arithmetical operations over fractions. Thus, there was accumulating material, which is added gradually to the most ancient mathematical direction, arithmetic. Measurement of space and volumes, the needs of construction equipment, and a little later, astronomy, cause the development of the beginnings of geometry."*

By comparing the views of the academicians Iliev (Bulgaria) and Kolmogorov (Russia) on the period of the origin of mathematics, it should be noted that these views mostly coincided to and were reduced to the following. At the stage of the origin of mathematics, two practical problems influenced the development of mathematics: the *counting* problem and the *measurement* problem.

The study of the *counting* problem ultimately led to the formation of such an important concept of mathematics as the *natural numbers* and to the creation of the *elementary number theory*, which solved important mathematical task in studying the properties of the natural numbers as well as solved the problem of creation of the *elementary arithmetic* that satisfied the needs of practice in performing the simplest arithmetic operations. The study of the problem of *measurement* led to the creation of *geometry*, and within this direction, the existence of the *irrational numbers*, the second most significant fundamental ancient mathematical discovery, which caused the *first crisis* in the foundations of mathematical science is proved.

By discussing the origins of mathematical science, we should not forget about another outstanding mathematical discovery of ancient mathematics, *Eudoxus' exhaustion method*, which, on the one hand, was created to overcome the *first crisis* in the foundation of mathematics, associated with the introduction of the *irrational numbers* into mathematics and, on the other hand, underlies the *Euclidean definition of* the *natural numbers*, which represents the same natural number as the sum of the "monads" $N = \underbrace{1 + 1 + \cdots + 1}_{N}$. It follows from these arguments that the *Eudoxus' exhaustion method* attempted to unite the two ancient problems that underlie the ancient mathematics: the problem of *counting*, which led to the *natural numbers*, and the problem of *measurement*, which led to the *irrational numbers*.

Volume II pursues two goals. The first goal is to set forth the foundation of the new mathematical measurement theory, the *Algorithmic Measurement Theory*, worked out by Alexey Stakhov in his doctoral dissertation, *Synthesis of Optimal Algorithms of Analog-to-Digital Conversion* (1972). The second goal is to show that this new mathematical theory of measurement is the foundation of all traditional positional numeral systems by starting with the *Babylonian positional numeral system* with base 60 and by ending with *decimal, binary, ternary*, and other traditional numeral systems. But the most important is that this new theory of measurement generates new, earlier unknown positional numeral systems, such as the *Fibonacci p-codes* [6, 16, 55, 56, 58, 60, 97], the *golden p-proportions codes* [6, 19, 53, 97], which are a generalization of the classical *binary system*, and finally, the *ternary mirror-symmetrical system and arithmetic* [72, 94, 97], which are a generalization of the classical *ternary system and arithmetic*.

Volume II consists of seven chapters, which can be divided into two parts. The first part includes Chapters 1 and 2.

Chapter 1 provides the introduction to the *Algorithmic Measurement Theory* [16], which is based on the *constructive approach* in modern mathematics.

Chapter 2 is devoted to the *Fibonacci measurement algorithms*, which generate the *Fibonacci p-codes* for mission-critical applications.

The second part of Volume II consists of five chapters (Chapters 3–7). Chapter 3 provides a brief statement of the most interesting facts in the history of traditional numeral systems (*Babylonian numeral system* with the base of 60, *Mayan numeral system, decimal, binary,* and *ternary* systems).

Chapter 4 is devoted to the description of the unique positional numeral system with irrational base $\Phi = \frac{1+\sqrt{5}}{2}$ (the *golden ratio*), proposed in 1957 by the American mathematician George Bergman [54], and following from Bergman's system the *"golden" number theory,* where the new properties of the natural numbers (the *Z- and D-properties*) are represented.

Chapter 5 is devoted to the description of the unique ternary arithmetic, the *"golden" ternary mirror-symmetrical arithmetic,* which opens the new direction in ternary computers.

Chapter 6 is devoted to a study of the *Fibonacci p-codes* and *Fibonacci arithmetic,* which are the new scientific results for computer science and can lead to designing of the *Fibonacci computers* for mission-critical applications.

Chapter 7 is devoted to the study of the general class of the redundant numeral systems. The classical binary system is the partial case of the codes of the golden $p$-proportions $(p = 0)$, the remaining "golden" codes, corresponding to the cases of $p = 1, 2, 3, \ldots$, are a generalization of *Bergman's system* $(p = 1)$, and for the general cases of $p = 1, 2, 3, \ldots$, they represent the general class of numeral systems with the irrational bases, which have a fundamental importance for mathematics (as the new definition of the real numbers) and also for computer science (as the basis of "golden" computers) and for the digital metrology (as the basis of the new theory of resistive dividers).

# About the Author

**Alexey Stakhov**, born in May 7, 1939, is a Ukrainian mathematician, inventor and engineer, who has made a contribution to the theory of Fibonacci numbers and the *golden section* and their applications in computer science and measurement theory. He is a Doctor of Computer Science (1972) and a Professor (1974), and the author of over 500 publications, 14 books and 65 international patents. He is also the author of many original publications in computer science and mathematics, including *algorithmic measurement theory* [16, 17], *Fibonacci codes and codes of the golden proportions* [19], *hyperbolic Fibonacci and Lucas functions* [64, 75] and finally the *Mathematics of Harmony* [6], which goes back in its origins to Euclid's *Elements*. In these areas, Alexey Stakhov has written many papers and books, which have been published in famous scientific journals by prestigious international publishers.

The making of Alexey Stakhov as a scientist is inextricably linked with the Kharkov Institute for Radio Electronics, where he was a postgraduate student of the Technical Cybernetics Department from 1963 to 1966. Here, he defended his PhD thesis in the field of Technical Cybernetics (1966) under the leadership of the prominent Ukrainian scientist Professor Alexander Volkov. In 1972, Stakhov defended (at the age of 32 years) his Grand Doctoral dissertation *Synthesis of Optimal Algorithms for Analog-to-Digital Conversion* (Computer Science speciality). Although the dissertation had an engineering character, Stakhov in his books and articles touched upon two fundamental problems of mathematics: *theory of measurement* and *numeral systems*.

Prof. Stakhov worked as "Visiting Professor" of different Universities: Vienna Technical University (Austria, 1976), University of Jena (Germany, 1986), Dresden Technical University (Germany, 1988), Al Fateh University (Tripoli, Libya, 1995–1997), Eduardo Mondlane University (Maputo, Mozambique, 1998–2000).

## Stakhov's Prizes and Awards

- Award for the best scientific publication by Ministry of Education and Science of Ukraine (1980);
- Barkhausen's Commemorative Medal issued by the Dresden Technical University as "Visiting Professor" of Heinrich Barkhausen's Department (1988);
- Emeritus Professor of Taganrog University of Radio Engineering (2004);
- The honorary title of "Knight of Arts and Sciences" (Russian Academy of Natural Sciences, 2009);
- The honorary title "Doctor of the Sacred Geometry in Mathematics" (American Society of the Golden Section, 2010);
- Awarded "Leonardo Fibonacci Commemorative Medal" (Interdisciplinary Journal "De Lapide Philosophorum", 2015).

# Acknowledgments

Alexey Stakhov expresses great thanks to his teacher, the outstanding Ukrainian scientist, Professor Alexander Volkov; under his scientific leadership, the author defended PhD dissertation (1966) and then DSc dissertation (1972). These dissertations were the first steps in Stakhov's research, which led him to the conceptions of *Mathematics of Harmony* and *Fibonacci computers* based on the *golden section* and *Fibonacci numbers*.

During his stormy scientific life, Stakhov met many fine people, who could understand and evaluate his enthusiasm and appreciate his scientific direction. About 50 years ago, Alexey Stakhov had read the remarkable brochure *Fibonacci Numbers* [8] written by the famous Soviet mathematician Nikolay Vorobyov. This brochure was the first mathematical work on, Fibonacci numbers published in the second half of the 20th century. This brochure, determined Stakhov's scientific interest in the *Fibonacci numbers* and the *golden section* for the rest of his life. In 1974, Professor Stakhov met with Professor Vorobyov in Leningrad (now St. Petersburg) and told Professor Vorobyov about his scientific achievements in this area. Professor Vorobyov gave Professor Stakhov, his brochure *Fibonacci Numbers* [8] as a keepsake with the following inscription: "*To highly respected Alexey Stakhov with Fibonacci's greetings*". This brief inscription because a certain kind of guiding star for Alexey Stakhov.

With deep gratitude, Stakhov recollects the meeting with the famous Austrian mathematician Professor *Alexander Aigner* in the Austrian city of Graz in 1976. The meeting with Professor Aigner was the beginning of the international recognition of Stakhov's scientific direction.

Another remarkable scientist, who had a great influence on Stakhov's research, was the Ukrainian mathematician and academician *Yuri Mitropolskiy,* the *Head of the Ukrainian Mathematical School* and the *Chief Editor of the Ukrainian Mathematical Journal.* His influence on Stakhov's researches, pertinent to the history of mathematics and other topics, such as the application of the Mathematics of Harmony in contemporary mathematics, computer science and mathematical education, were inestimable stimulus for Alexey Stakhov. Thanks to the support of Yuri Mitropolskiy, Stakhov published many important articles in the prestigious Ukrainian academic journals, including the *Ukrainian Mathematical Journal.*

In 2002, *The Computer Journal* (British Computer Society) published the fundamental article by Stakhov, *"Brousentsov's Ternary Principle, Bergman's Number System and Ternary Mirror-Symmetrical Arithmetic"* (*The Computer Journal,* Vol. 45, No. 2, 2002) [72]. This article by Stakhov created great interest among all the English scientific computer community. Emeritus Professor of Stanford University *Donald Knuth* was the first outstanding world scientist, who congratulated Prof. Stakhov with this publication. Donald Knuth's letter became one of the main stimulating factors for writing Stakhov's future book *The Mathematics of Harmony From Euclid to Contemporary Mathematics and Computer Science* (World Scientific, 2009) [6]. *Professor Stakhov considers this book as the main book of his scientific life.*

Stakhov's arrival to Canada in 2004 became the beginning of a new stage in his scientific research. Within 15 years (2004–2019), Prof. Stakhov had published more than 50 fundamental articles in different international English-language journals, such as *Chaos, Solitons & Fractals, Applied Mathematics, Arc Combinatoria, The Computer Journal, British Journal of Mathematics and*

*Computer Science, Physical Science International Journal, Visual Mathematics*, etc. Thanks to the support of Prof. El-Nashie, the Editor-in-Chief of the Journal *Chaos, Solitons & Fractals* (UK), Stakhov published in this journal 15 fundamental scientific articles that garnered great interest among the English-speaking scientific community.

The publication of the three fundamental books *The Mathematics of Harmony* (World Scientific, 2009) [6], *The "Golden" Non-Euclidean Geometry* (World Scientific, 2016, co-author Prof. Samuil Aanson) [52] and *Numeral Systems with Irrational Bases for Mission-Critical Applications* (World Scientific, 2017) [53] is one of the main scientific achievements by Stakhov during the Canadian period of his scientific creativity. These books were published thanks to the support of the famous American mathematician Prof. *Louis Kauffman*, Editor-in-Chief of the *Series on Knots and Everything* (World Scientific) and Prof. *M.S. Wong*, the famous Canadian mathematician (York University) and Editor-in-Chief of the *Series on Analysis, Application and Computation* (World Scientific). A huge assistance in the publication of Stakhov's books by of World Scientific was rendered by the American researcher, Scott Olsen, Professor of Philosophy at the College of Central Florida, and Jay Kappraff, Emeritus Professor of Mathematics at the New Jersey Institute of Technology. Prof. Scott Olsen, who was one of the leading US experts in the field of *Harmony* and the *golden section*, was the English editor for Stakhov's book mentioned above and the Emeritus Professor Jay Kappraff was the reviewer of Stakhov's book, *The Mathematics of Harmony* (World Scientific, 2009).

The prominent Ukrainian mathematician and head of the Ukrainian Mathematical School, *Yuri Mitropolskiy*, praised highly Stakhov's *Mathematics of Harmony*. Academician Mitropolsky organized Stakhov's speech at the meeting of the Ukrainian Mathematical Society in 1998. Based upon his recommendation, Stakhov's articles were published in the Ukrainian academic journals, in particular, the *Ukrainian Mathematical Journal*. Under his direct influence, Stakhov started writing the book, *The Mathematics of Harmony. From Euclid to Contemporary Mathematics and Computer*

*Science* [6], which was published by World Scientific in 2009 following the death of the academician Mitropolsky in 2008.

## Scientific cooperation of Alexey Stakhov and Samuil Aranson

Samuil Aranson's acquaintance to the *golden section* and the *Fibonacci numbers* began in 2001 after the reading of a very rare book "Chain Fractions" [107] by the famous Russian mathematician, Aleksandr Khinchin. In this book, Samuil Aranson found results, related to the representation of the *"golden ratio"* in the form of a continued fraction.

In 2007, Prof. Aranson read a wonderful Internet publication, *Museum of Harmony and Golden Section*, posted in 2001 by Professor Alexey Stakhov and his daughter Anna Sluchenkova. This Internet Museum covers various areas of modern natural sciences and tells about the different and latest scientific discoveries, based on the *golden ratio* and *Fibonacci numbers*, including the *Mathematics of Harmony* and its applications in modern natural sciences. After reading this Internet Museum, Samuil Aranson contacted Alexey Stakhov in 2007 through e-mail and offered him joint scientific collaboration in further application of the *Mathematics of Harmony* in various areas of mathematics and modern natural sciences. Scientific collaboration between Alexey Stakhov and Samuil Aranson turned out to be very fruitful and continues up to the present time.

New ideas in the field of elementary mathematics and the history of mathematics, developed by Stakhov (*Proclus's hypothesis* as a new look at Euclid's *Elements* and history of mathematics, *hyperbolic Fibonacci and Lucas functions* [64, 75] as a new class of elementary functions and other mathematical results) attracted the special attention of Prof. Aranson. Scientific collaboration between Stakhov and Aranson began in 2007. From 2007, they published the following joint scientific works (in Russian and English), giving fundamental importance for the development of mathematics and modern theoretical natural sciences:

## Stakhov and Aranson's Mathematical Monographs in English

1. Stakhov A., Aranson S., *The Mathematics of Harmony and Hilbert's Fourth Problem. The Way to the Harmonic Hyperbolic and Spherical Worlds of Nature*. Germany: Lambert Academic Publishing, 2014.
2. Stakhov A., Aranson S., Assisted by Scott Olsen, *The "Golden" Non-Euclidean Geometry: Hilbert's Fourth Problem, "Golden" Dynamical Systems, and the Fine-Structure Constant*, World Scientific, 2016.


## Stakhov and Aranson's Scientific Papers in English

3. Stakhov A.P., Aranson S.Kh., "Golden" Fibonacci goniometry, Fibonacci-Lorentz transformations, and Hilbert's fourth problem. *Congressus Numerantium* **193**, (2008).
4. Stakhov A.P., Aranson S.Kh., Hyperbolic Fibonacci and Lucas functions, "golden" Fibonacci goniometry, Bodnar's geometry, and Hilbert's fourth problem. Part I. Hyperbolic Fibonacci and Lucas functions and "Golden" Fibonacci goniometry. *Applied Mathematics* **2**(1), (2011).
5. Stakhov A.P., Aranson S.Kh., Hyperbolic Fibonacci and Lucas functions, "golden" Fibonacci goniometry, Bodnar's geometry, and Hilbert's fourth problem. Part II. A new geometric theory of phyllotaxis (Bodnar's Geometry). *Applied Mathematics* **2**(2), (2011).
6. Stakhov A.P., Aranson S.Kh., Hyperbolic Fibonacci and Lucas functions, "golden" Fibonacci goniometry, Bodnar's geometry, and Hilbert's fourth problem. Part III. An original solution of Hilbert's fourth problem. *Applied Mathematics* **2**(3), (2011).
7. Stakhov A.P., Aranson S.Kh., The mathematics of harmony, Hilbert's fourth problem and Lobachevski's new geometries for physical world. *Journal of Applied Mathematics and Physics* **2**(7), (2014).

8. Stakhov A., Aranson S., The fine-structure constant as the physical-mathematical millennium problem. *Physical Science International Journal* **9**(1), (2016).

9. Stakhov A., Aranson S., Hilbert's fourth problem as a possible candidate on the millennium  problem in geometry. *British Journal of Mathematics & Computer Science* **12**(4), (2016).

# Chapter 1

# Foundations of the Constructive (Algorithmic) Measurement Theory

## 1.1. The Evolution of the Concept of "Measurement" in Mathematics

What is a "measurement"? In the Great Soviet Encyclopedia, we find the following definition of this concept:

> *"Measurement is the operation, by means of which the ratio of one (measurable) magnitude to another homogeneous magnitude (taken as the unit of measurement) is determined; the number, determined with such ratio, is called the numerical value of the measured magnitude".*

Measurement is an important way of the quantitative cognition of the objective world. The Great Russian scientist, Dmitry Mendeleev, the creator of the Periodic Table of Chemical Elements, expressed his opinion about the *measurement* as follows:

> *"Science begins from the "measurement". Exact science is unthinkable without measure".*

The problem of measurement plays the same significant role in mathematics as in other areas of science, in particular, engineering, physics, and other "exact" sciences.

Let's now trace the evolution of the concept of "measurement" in mathematics [16]. As is known, the first "measurement theory"

1

was the set of rules, which had been used by the ancient Egyptian surveyors. From this set of rules, as the ancient Greeks testify, *geometry* was obligated with its appearance (and name) to the problem of the "Earth's measurement".

However, already in Ancient Greece, the measurement problems had been divided into applied tasks related to *logistics* and fundamental problems related to *geometry*; the ancient mathematics focussed on the latter. The science of measurement was developing during this period primarily as a *mathematical theory*.

The discovery of *incommensurable segments*, made in the scientific school of Pythagoras, was one of the main mathematical achievements of this period. This discovery caused the first crisis in the foundations of mathematics and led to the introduction of *irrational numbers* — the second (after *natural numbers*) fundamental concept of mathematics. This discovery led to the formulation of the *Eudoxus method of exhaustion* and *measurement axioms* (see Section 1.2.), to which theory of numbers, integral, and differential calculus go back in their origins.

## 1.2. Axioms of Eudoxus–Archimedes and Cantor

To overcome the first crisis in the foundations of ancient mathematics, associated with the discovery of "incommensurable segments", the outstanding geometer Eudoxus proposed the "method of exhaustion", by which he built the ingenious theory of relations, underlying the ancient theory of the continuum. "Method of exhaustion" played a prominent role in the development of mathematics. Being a prototype of integral calculus, the "exhaustion method" allowed ancient mathematicians to solve the problems of calculating volumes of a pyramid, a cone, a ball. In modern mathematics, the "method of exhaustion" is reflected in the *Eudoxus–Archimedes axiom*, also called the *measurement axiom*.

The theory of measuring geometric quantities, which goes back to "incommensurable segments", is based on a group of axioms, called continuity axioms [16], which include the Eudoxus–Archimedes and Cantor axioms or the Dedekind axioms.

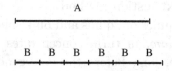

Fig. 1.1. Eudoxus–Archimedes axiom.

A₀  A₁   A₂  Aₙ  Bₙ  B₂   B₁  B₀

C

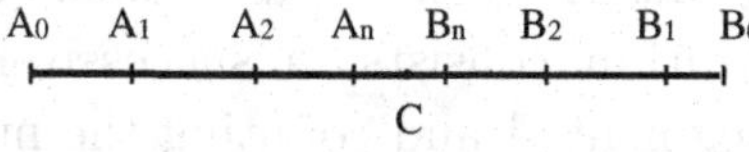

Fig. 1.2. Cantor's axiom.

**Eudoxus–Archimedes axiom (the "measurement axiom").**
*For any two segments A and B (Fig. 1.1), we can find the positive integer n such that*

$$nB > A. \qquad (1.1)$$

**Cantor's axiom (about the "contracted segments").** *If an infinite sequence of segments $A_0B_0$ , $A_1B_1$, $A_2B_2$, ..., $A_nB_n$, ... "nested" into each other is given (Fig. 1.2), that is, when each segment is part of the previous one, then there is at least one point of C, common to all segments.*

The main result of the theory of the geometric quantities is the proof of the existence and uniqueness of the solution $q$ of the *basic measurement equation*:

$$Q = qV, \qquad (1.2)$$

where $V$ is the unit of measurement, $Q$ is the measurable quantity, and $q$ is the result of measurement.

Despite the seeming simplicity of the axioms, formulated above, and the whole mathematical theory of measurement, it is nonetheless the result of more than 2000 years in the development of mathematics and contains a number of in-depth mathematical ideas and concepts.

First of all, it is necessary to emphasize that the *Eudoxus exhaustion method* and the *measurement axiom*, resulting from it (Fig. 1.1), are of practical (empirical) origin; they were borrowed by the ancient Greek mathematicians in the practice of measurement.

In particular, the "exhaustion method" is the mathematical model for measuring the volumes of liquids and bulk solids by "exhaustion"; the *measurement axiom*, in turn, concentrates thousands of years of human experience, long before the emergence of the axiomatic method in mathematics, billions of times measuring distances, areas and time intervals, and is a concise formulation of an algorithm of measuring line segment $A$ by using a segment $B$ (Fig. 1.1). The essence of this algorithm consists in successively postponing the segment $B$ on the segment $A$ and counting the number of segments $B$ that fit on the segment $A$. In modern measurement practice, this measurement algorithm is called the *counting algorithm*.

### 1.3. The Problem of Infinity in Mathematics

Cantor's axiom (Fig. 1.2) contains another amazing creation of mathematical thought, an abstraction of *actual infinity*. It is this idea of infinity that underlies the Cantor theory of the infinite sets [16].

The idea of *actual infinity* as the main idea of Cantor's (set-theoretic) style of mathematical thinking was strongly criticized by the representatives of the so-called *completed one it is impossible without gross violence over the mind, which rejects such contradictory fantasies.*

Earlier, the famous mathematician David Hilbert (known for his "finite" installations) expressed the same idea in other words and, by discussing the concepts of the *finite* and *infinite*, came to the following conclusion [125]:

> *"From all our reasoning, we want to make some summary on infinity; the general conclusion is the following: the infinite is not realizing anywhere. The infinite does not exist in Nature, and this concept is unacceptable as the basis of our rational thinking — here we have a wonderful harmony between being and thinking ... Operation with the infinite can become reliable only through the finite".*

The paradoxes or contradictions in Cantor's theory of infinite sets, discovered at the beginning of the 20th century, significantly shook the foundations of mathematics. Various attempts have been made to strengthen them. The most radical of them is the constructive direction in the substantiation of mathematics [124], which completely excludes the consideration of the abstraction of actual infinity and uses a much more "modest" abstraction of the infinite called the *abstraction of potential feasibility*.

The contradiction between potential and actual infinities manifests itself most vividly in the mathematical theory of measurement [16], when we analyze the *Eudoxus–Archimedes axioms* (Fig. 1.1) and the *Cantor axioms* (Fig. 1.2).

To clarify this contradiction, let's consider once again the "basic measurement equation" (1.2). The idea of this contradiction is as follows. With the help of the *Eudoxus–Archimedes axiom*, a certain sequence of the "contractible" segments, which are compared with the measured segment $Q$, is formed from the measurement unit $V$ according to certain rules, called the *measurement algorithm*; when this process strives to infinity, on the basis of *Cantor's axiom*, for any $Q$ and given $V$, such segment $Q$, formed from $V$, which "absolutely exactly" coincide with $Q$, will always be found. The most essential part of this proof is a presentation about measurement, as a process that ends in an infinite time (according to Cantor's axiom). Thus, at the initial stage of the proof of equation (1.2), we use the concept of the *potential infinity* (Eudoxus–Archimedes axiom), and at the final stage, we "jump over" this concept and use the concept of the *actual infinity* (Cantor's axiom). In this regard, it is appropriate to cite the following quotation from the book [16]:

> "It is appropriate to pay attention to the internal contradictoriness (in the dialectical sense) of the set-theoretic theory of measurement (and as a consequence of the theory of real numbers), which allows in its initial positions (continuity axiom) the coexistence of dialectically contradictory ideas about actual infinity: the actual infinityin Cantor's axiom (and Dedekind's axiom) and the potential infinity, that is, "becoming", unfinished in Archimedes' axiom".

The existing mathematical theory of measurement and the theory of real numbers arising from it, based on *Cantor's axiom*, are internally contradictory; but such theories, based on the contradictory axioms, cannot be the basis for mathematical reasoning! Otherwise, all mathematics becomes an internally contradictory theory. One wonder what had actually happened in mathematics at the beginning of the 20th century, when the contradictions were found in *Cantor's theory* of infinite sets. It is surprising that such a simple idea was not noticed by mathematicians before the book [16].

## 1.4. Criticism of the Cantor Theory of Infinite Sets

### 1.4.1. Infinitum Actu Non Datur

As it is well known, mathematics became a deductive science in ancient Greece. Already in 6th century BC, Greek philosophers studied the problem of *infinity* and the *continuous* and *discrete* problems related to it. Aristotle paid much attention to the development of this concept. He was the first who categorically began to object against the application of the actual infinity in science, referring to the fact that, despite knowing the methods of counting the finite number of objects, we cannot use the same methods to infinite sets. In his *Physics*, Aristotle stated as follows:

> *"There remains the alternative, according to which the infinite has only potential existence ... The actual infinite is not exist."*

According to Aristotle, mathematics does not need *actual infinity*. Aristotle is the author of the famous thesis *Infinitum Actu Non Datur*, which translates from Latin as the statement about the impossibility of the existence of the logical or mathematical (that is, only imaginable, but not existing in Nature) actual-infinite objects.

### 1.4.2. Criticism of Cantor's theory of sets in 19th and early 20th centuries

Cantor's theory of infinite sets caused a storm of protests already in the 19th century. The detailed analysis of the criticism of this theory was given in Chapter IX "Paradise Barred: A New Crisis of Reason" of the remarkable book by the American historian of mathematics Morris Kline *Mathematics. Loss of Certainty* [101].

Many famous mathematicians of the 19th century spoke out sharply negatively about this theory. Leonid Kronecker (1823–1891), who had a personal dislike to Cantor, called him a *charlatan*. Henri Poincaré (1854–1912) called Cantor's theory of the infinite sets the *serious illness* and considered it as a kind of *mathematical pathology*. In 1908, he declared as follows:

> *"The coming generations will regard the theory of sets as the disease, from which they have recovered."*

Unfortunately, Cantor's theory had not only opponents but also supporters among famous scientists and thinkers. Russell called Cantor one of the great thinkers of the 19th century. In 1910, Russell wrote: *"Solving problems that have long enveloped mystery in mathematical infinity is probably the greatest achievement that our age should be proud of."* In his speech at the First International Congress of Mathematicians in Zurich (1897), the famous mathematician Hadamard (1865–1963) emphasized that the main attractive feature of Cantor's infinite set theory is that for the first time in mathematical history, the classification of the sets, based on the concept of the *cardinal number*, was given. In his opinion, the amazing mathematical results, which follow from Cantor's set theory, should inspire mathematicians to new discoveries. Thus, in Hadamard's speech, Cantor's theory of infinite sets was elevated to the level of the main mathematical theory, which can become the foundation of mathematics.

### 1.4.3. Research by Alexander Zenkin

In the recent years, in the works of the outstanding Russian mathematician and philosopher Alexander Zenkin [126], as well as in the works of other authors [127–129], radical attempts have been made to "purify" mathematics from Cantor's theory of sets based on conception of actual infinity (Fig. 1.3).

Fig. 1.3.  Alexey Stakhov and Alexander Zenkin.

(Moscow University, April 29, 2003: Stakhov's lecture "A New Type of the Elementary Mathematics and Computer Science Based on the Golden Section", delivered at the joint meeting of the seminar *Geometry and Physics*, Department of Theoretical Physics, Moscow University, and Interdisciplinary Seminar *Symmetries in Science and Art* of the Institute of Mechanical Engineering, Russian Academy of Sciences.)

The analysis of the *Cantor theory of infinite sets*, presented in [126], led Alexander Zenkin to the conclusion that the proofs of many of Cantor's theorems are *logically incorrect*, and the *Cantor theory of the infinite sets* is in a certain sense the *mathematical hoax of the 19th century*. Some famous mathematicians of the 19th century were fascinated by *Cantor's theory* and, by accepting his unusual theory without proper critical analysis, elevated it to the rank of the greatest mathematical discovery of the 19th century and laid the foundations of mathematics.

The discovery of the paradoxes in the *Cantor theory of the infinite sets* considerably cooled the enthusiasm of mathematicians toward this theory. Alexander Zenkin [126] put the final point in the critical analysis of *Cantor's theory*. He showed that the *main Cantor error* was the adoption of abstraction of the *actual infinity*, which, according to Aristotle, is unacceptable in mathematics.

But without the abstraction of *actual infinity*, Cantor's theory of infinite sets is untenable! As mentioned above, Aristotle was the first scientist and thinker who drew attention to this problem and warned about the impossibility of using the concept of the *actual infinity* in mathematics ("Infinitum Actu Non Datur").

In Stakhov's article [129], the following question was posed: *"Is modern mathematics not standing on the "pseudoscientific" foundation?"* So far, mathematicians have not answered this question concerning the foundations of mathematics.

## 1.5. Constructive Approach to the Creation of the Mathematical Measurement Theory

In the framework of the *constructive approach* to the creation of the mathematical measurement theory, the concept of the *actual infinity* should be excluded from the consideration due to its internal contradictoriness (the "completed infinity").

Modern constructivist mathematicians, who consider the abstraction of *actual infinity* as an *internally contradictory concept* (the *completed infinity*), came up with two ideas that can become the basis of the *constructive (algorithmic) theory of measurement* [16]:

(1) An intuitive, "practical" idea about the *finiteness* of the measurement process, according to which every measurement is completed in the finite number of steps.
(2) Constructive idea of the *potential feasibility*, in accordance to which we ignore the limitations of our possibilities in choosing the measuring means and the number of measurement steps (i.e., the number of measurement steps is always *finite* and can be choosen to be arbitrarily large and there exists always the *potential possibility* to do the next measurement step).

Such a seemingly insignificant change in the approach to the measurement leads us to rethink many problems of the mathematical theory of measurement. With the set-theoretic approach, the measurement is carried out to the "point", i.e., to the absolutely exact coincidence of the measurable and measuring segments (a possibility of such an absolutely accurate measurement follows from *Cantor's axiom*).

With the *constructive approach*, the measurement never reaches the "point", and the measurement result reduces always to certain *uncertainty interval* regarding the true value of the measurable quantity. By increasing the number of measurement steps, this interval narrows and can be made arbitrarily small, but this interval never turns into the "point".

In his famous work *On the Philosophy of Mathematics* [130], Hermann Weil pays attention to the following distinction between the classical and constructive definitions of the *continuum* concept:

> "*In modern analysis, the continuum is considered as the set of its points; in the continuum, it sees only a special case of the basic logical relationship between an element and a set. But it is amazing that the same fundamental relationship between the whole and its part had not yet found a place in mathematics! Meanwhile, the possession of parts is the basic property of the continuum and Brauer's theory puts this relationship in the basis of the mathematical study of the continuum. This is actually the basis of the above attempt to proceed not from points, but from intervals, as from the primary elements of the continuum*".

One of the important moments in the set-theoretic theory of measurement, based on the *Eudoxus–Archimedes axiom* and the *Cantor axiom*, is the choice of the *measurement algorithm*, which is defined by the numeral system, in which the *measurement result* is represented.

With the infinite (in terms of the *actual infinity*) number of measurement steps, i.e., when we are measuring up to the "point", the *measurement algorithm* does not affect the final measurement result and therefore the problem of choosing *measurement algorithms* as a serious mathematical problem does not arise here. The choice of the measurement algorithm has an arbitrary character, and, as a rule, it reduces to the "decimal" or "binary" algorithms.

With the infinite (in terms of the *potential infinity*) number of measurement steps, i.e., when we are measuring up to the "interval" [130], between the measurement algorithms, there arises the difference in the measurement "accuracy", achieved with the help of the given measurement algorithm. Let's recall that the *measurement "accuracy"* is equal to the ratio of the initial uncertainty interval to the uncertainty interval on the final step of measurement. For such conditions, the constructive idea about the "efficiency" of the *measurement algorithms* [131] comes into play, and the problem of the synthesis of *efficient* or *optimal* measurement algorithms is put forward as the central problem of the constructive (algorithmic) theory of measurement [16].

Thus, the constructive approach to the theory of measurement leads us to the formulation of the problem, which, in essence, has never been considered in the mathematical theory of measurement as a serious mathematical problem, namely the problem of finding the *"optimal" measurement algorithms*. The solution of such problem led us to the creation of the *algorithmic measurement theory* [16], which can be considered as the *constructive direction* in the mathematical theory of measurement.

## 1.6. The "Indicatory" Model of Measurement

### 1.6.1. The conceptions of the "indicatory" element (IE) and the "indicatory" model of measurement

When we set forth the task of creating the *algorithmic measurement theory*, it is necessary to clarify once again what the measurement is, what its purpose is, what the *measurement algorithm* is and what tools we use to implement the *measurement*.

First of all, we note that if we want to measure something, we must know some *source data* concerning the measurement, in particular, the *range of measurable values*. It is one thing to measure

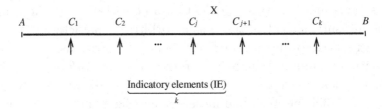

Fig. 1.4. Constructive ("indicatory") model of measurement.

the cosmic distances, for example, the distance from the Earth to the Sun and quite another to measure the atomic distances.

However, when we turn to the *mathematical measurement*, we are distracted from the physical character of the *measured quantities*; in this case, for all the cases of *measurement*, we can represent the *measurement range* in the form of the *geometric segment AB*.

It is clear that the *measurable value* is one of the possible values, belonging to this range, that is, before the beginning of the measurement, there is some *uncertainty* regarding the *measurable value*; otherwise, the measurement would simply be *meaningless*. This situation of the "uncertainty" can be depicted by using the "unknown" *point X* located on the segment *AB* (Fig. 1.4).

Now, we can formulate the purpose of *measurement*. The purpose of measurement consists in the determination of the length of the segment *AX* (Fig. 1.4). In practice, this purpose is realized with the help of special devices, for example, the "lever scales" or "comparators".

The "comparators" carry out the comparison of measurable quantity with certain "standard quantities" or "measures", formed from the "measurement unit" *V* and, depending on the result of the comparison, they give us information about the position of the measured segment *AX* on the initial segment *AB*. Thus, the essence of the measurement reduces to the successive comparisons of the measurable segment *AX* with some "measures", which are formed at each measurement step from the unit *V* according to the measurement algorithm.

In order to model the process of the comparison of the *measurable segment AX* to the "measures", the important concept of "indicatory element" (IE), which is a peculiar model of the "comparator" or the "lever scales", the basic means of any *measurement*, was introduced in Ref. [16].

We assume that each "indicatory element" (IE) can be enclosed to some "known" point $C$ of the segment $AB$, according to the *measurement algorithm*, that is, we can form for every "indicatory element" some "measure" on each step of the measurement.

The "indicatory element" gives us information about the relative position of the "unknown" point $X$ and the "known" *point $C$*. If the IE is to the right of the point $X$, it "indicates" the binary signal 0; otherwise, the binary signal 1.

## 1.7. The Concept of the Optimal Measurement Algorithm

In Section 1.6, we construct the following *mathematical model* of *measurement* (see Fig. 1.4), which allows solving the *task of the synthesis of the optimal measurement algorithms*.

The essence of this task consists in the following. Suppose that on the segment $AB$, there is the "unknown" *point $X$*. The length of the segment $AX$ is determined by using $k$ "indicatory elements" (see Fig. 1.4).

As a result of the enclosing of $j$th IE $(j = 1, 2, \ldots, k)$ in the first step of the algorithm to the point $C_j (j = 1, 2, \ldots, j, j + 1, \ldots, k)$, the "comparison" of the segment $AX$ to the segment $AC_j (j = 1, 2, \ldots, j, j + 1, \ldots, k)$ is realized, i.e., the relations of the "less" $(<)$ or the "greater than or equal to" $(\geq)$ between comparable segments $AX$ and $AC_j$ $(j = 1, 2, \ldots, j, j + 1, \ldots, k)$ are determined. Note that the indicated property of the IE is its main definition.

We assume that the "output signal" or "indication" of the $j$th IE at the point $C_j (j = 1, 2, \ldots, j, j + 1, \ldots, k)$ takes the value 0 if the relation $AX < AC_j$ holds, and the value 1 if $AX \geq AC_j$. We will also assume that the $j$th IE is "to the right" relative to the point $X$, if as a result of its enclosing to the certain point $C_j$ $(j = 1, 2, \ldots, j, j + 1, \ldots, k)$, its "indication" signal takes the value 0, and is "to the left" (otherwise the "indication" signal takes the value 1).

Note that the task of measuring the segment $AX$ in the "indicatory" model of the measurement (Fig. 1.4) reduces to narrowing the *interval of uncertainty* about the point $X$. The process of *measurement* consists in the fact that, at the first step of the measurement, the "indicatory elements" (IE) are enclosed to some points $C_j(j = 1, 2, \ldots, j, j + 1, \ldots, k)$ of the *initial uncertainty interval $AB$* and, on the basis of the "indication" *signals* of the IE in these points, the *uncertainty interval* narrows down to some new segment $A_1 B_1 \subset AB$; in all the subsequent steps, the "indicatory elements" enclose to the points of the *uncertainty interval $A_1 B_1$* chosen at the previous step.

Some conditions or *restrictions $S$* can be imposed on the measurement process. The system of formal rules, which determines for each segment $AX$ at each of the $n$ measurement steps the set of the points $C_j(j = 1, 2, \ldots, j, j + 1, \ldots, k)$ of enclosing the $k$ "indicatory elements", based on their "indications" on the previous steps for the *restrictions $S$*, is called the $(n, k, S)$-algorithm.

Thus, according to the above constructive "indicatory" model of *measurement* (Fig. 1.4), the measurement process consists in the sequential enclosing of the IE to some points $C_j(j = 1, 2, \ldots, j, j + 1, \ldots, k)$ of the segment $AB$ and narrowing the uncertainty interval regarding the point $X$.

The constructed model of measurement reduces the task of measuring the measurable segment $AX$ to the task of deterministic one-dimensional search of the coordinate of the point $X$ on the segment $AB$ by using $k$ IE in $n$ steps. Such a method makes it possible to use the ideas and principles of the *optimum seeking methods* [131] to the strict definition of the *optimal $(n, k, S)$-algorithm*.

As a result, the action of the $(n, k, S)$-algorithm on the segment $AB$ is the certain interval of the uncertainty $\Delta$, containing the point $X$ at the last step of the algorithm. The length of this interval $\Delta$ determines the "exactness" of measuring the segment $AX$. By considering the operation of the $(n, k, S)$-algorithm for all possible points $X$ of the segment $AB$ and by highlighting all the intervals of uncertainty, that contain the corresponding points of $X$, we get the set of *the uncertainty intervals*:

$$\Delta_1, \Delta_2, \ldots, \Delta_N, \qquad (1.3)$$

which form the *partition* of the segment $AB$ into $N$ segments, wherein

$$AB = \Delta_1 + \Delta_2 + \cdots + \Delta_N. \qquad (1.4)$$

It is clear that the partition (1.3), satisfying the relation (1.4), is an "integral" characteristic of the *effectivenes* of the $(n, k, S)$-algorithm, acting on the segment $AB$; therefore, it can be used to introduce the strict definition of the notion of the *optimal* $(n, k, S)$-algorithm.

Let's consider the ratio $T_i = \frac{AB}{\Delta_i}$ $(i = 1, 2, 3, \ldots, N)$, which will be called the *exactness* of determining the point $X$, belonging to the interval $\Delta_i$. Let's select from the *partition* (1.3) the largest interval of uncertainty $\Delta_{\max}$. The segment $\Delta_{\max}$ corresponds to the smallest (i.e., the "worst") *exactness* of determining the point $X$

$$T_{\min} = \frac{AB}{\Delta_{\max}}, \qquad (1.5)$$

with the help of which we will evaluate the *effectiveness* of the operation of the $(n, k, S)$-algorithm on the segment $AB$. It is clear that each *partition* (1.3) is characterized by the only value of (1.5). We call this value, defined by the expression (1.5), the $(n, k)$-*exactness* of the $(n, k, S)$-algorithm.

By using the concept of the $(n,k)$-*exactness* (1.5) as the criterion of the "effectiveness" or "optimality" of the $(n, k, S)$-algorithms (for the given values $n, k$, and the condition $S$), we can compare different $(n, k, S)$-algorithms with respect to their $(n, k)$-*exactness*.

The above consideration gives us a possibility introducing the following rigorous definition of the *optimal* $(n, k, S)$-*algorithm*.

**Definition 1.1.** The $(n, k, S)$-algorithm is called optimal if, for all other equal conditions, that is, for the given parameters of $n, k, S$, this algorithm provides the highest value of the $(n, k)$-exactness (1.5).

Note that this definition essentially uses the so-called *mini–max principle*, which is widely used in the modern theory of the *optimal systems* [131]. According to this principle, we call a strategy, which ensures the *maximum value* of the criterion of the *effectiveness* (1.5) for the *worst case*, as "optimal".

Let's consider once again the partition (1.3), which is the result of the operation of the $(n, k, S)$-algorithm on the segment $AB$, and let's estimate this partition from the point of view of the concept of the $(n, k)$-exactness (1.5) introduced above. It is easy to show that the highest value of the $(n, k)$-*exactness* (1.5), for other equal conditions, corresponds to the *uniform partition* of the segment, when all the intervals of the uncertainty (1.3) are equal to each other, that is,

$$\Delta_1 = \Delta_2 = \ldots = \Delta_N = \Delta_{\max} = \frac{AB}{N}. \tag{1.6}$$

By substituting in (1.5) instead $\Delta_{\max}$ its value, given by (1.6), we obtain the following expression for the $(n, k)$-*exactness* of the *measurement algorithm*:

$$T = T_{\min} = N, \tag{1.7}$$

where $N$ is the number of the *quantization levels*, ensured by the *optimal* $(n, k, S)$-algorithm. This means that the *optimal* $(n, k, S)$-algorithm always provides the *uniform partition* of the segment $AB$; at the same time, the *optimal* algorithm is an $(n, k, S)$-algorithm that provides (for the given parameters $n$, $k$, $S$ of the algorithm) the partition of the initial segment $AB$ into the most number $N$ of equal intervals.

It is clear that for the given *restriction $S$*, the efficiency criterion of the $(n, k, S)$-algorithm, that is, the number of *quantization levels $N$*, depends on the number of steps $n$ of the algorithm and the number of "*indicatory elements*" $k$, involved in the *measurement*, i.e.,

$$N = F(n, k). \tag{1.8}$$

Further, the function $F(n, k)$ will also be called the *efficiency function* of the *measurement algorithm*.

## 1.8. Classical Measurement Algorithms

In the modern measurement technology (in particular, in the technique of analog-to-digital conversion), the following *measurement algorithms* are widely used:

### 1.8.1. Counting algorithm

This algorithm uses only one IE ($k = 1$) and is implemented in the $n$ steps; at the same time, the segment $AB$ is divided into $n + 1$ equal parts in the $n$ steps, i.e., in this case, the $(n, 1)$-exactness of the $(n, 1, S)$-algorithm is determined by the following *efficiency function*:

$$N = F(n, 1) = n + 1. \tag{1.9}$$

It is important to emphasize that this *measurement algorithm* takes its origin in measurement praxis and has deep roots in *ancient mathematics*. It is this algorithm that underlies the *Eudoxus exhaustion method* and the *Euclidean definition* of the *natural number*:

$$N = \frac{1 + 1 + \ldots + 1}{N}, \tag{1.10}$$

which defines not only the *natural numbers* but also all the problems of the *elementary theory of numbers*, the foundations of which are set out in Euclid's *Elements*.

### 1.8.2. "Binary" algorithm

This algorithm also uses the only one IE ($k = 1$) and is implemented in the $n$ steps; at the same time, the initial segment $AB$ is divided into the $2^n$ equal parts, i.e., in this case, the $(n, 1)$-exactness of the algorithm is determined by the following *efficiency function*:

$$N = F(n, 1) = 2^n. \tag{1.11}$$

Note that this algorithm "generates" the binary representation of integers (the "binary system"), which underlies the modern

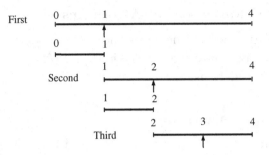

Fig. 1.5. The 3-step "counting algorithm".

information technology:

$$N = \sum_{i=0}^{n-1} a_i 2^i, \tag{1.12}$$

where $a_i \in \{0, 1\}$ is the bit and $2^i (i = 0, 1, 2, \ldots, n-1)$ is the weight of the $i$th digit.

Thus, the above classic measurement algorithms (*counting algorithm* and *"binary" algorithm*), on the one hand, generate *elementary theory of numbers* (*counting algorithm*) and on the other hand, are the *basis of modern computer science* (*"binary" algorithm*).

### 1.8.3. Readout algorithm

This algorithm is realized in one step $(n = 1)$ and uses $k$ IE; at the same time, by using $k$ IE, the segment $AB$ is divided into $k+1$ equal parts, i.e., in this case, the $(1, k)$-exactness of the algorithm is determined by the following *efficiency function*:

$$N = F(1, n) = k + 1. \tag{1.13}$$

As an example, let's consider more in detail the operation of the classic measurement algorithms: the *counting algorithm* (Fig. 1.5), the *"binary" algorithm* (Fig. 1.6) and the *readout algorithm* (Fig. 1.7).

The *counting algorithm*, shown in Fig. 1.5, is realized in three steps and uses only one IE $(k = 1)$. It divides the segment $[0, 4]$ into four equal parts.

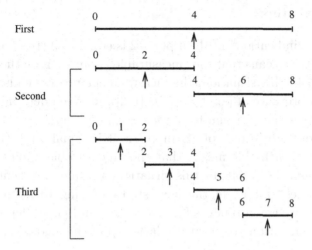

Fig. 1.6. The "binary" measurement algorithm.

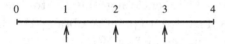

Fig. 1.7. Readout algorithm.

**The first step** consists in the enclosing IE to point 1. In this case, depending on the *indication* of the IE, two situations may arise: [0,1] and [1,4].

**The second step**:

(a) If the "indication" of IE at point 1 is equal to 0 (IE shows *to the left*), this means that the "measurable" point $X$ is on the segment [0,1]. In this situation, the measurement process ends, because the coordinate of point $X$ ($X \in [0,1]$) has been defined with "exactness" equal to the *measurement unit*.

(b) If the "indication" of IE at point 1 is equal to 1 (IE shows *to the right*), this means that the "measurable" point $X$ is located on the interval [1,4]. In this situation, the measurement process continues and IE in the next step is enclosing to point 2; as a result, we get two new situations: [1,2] and [2,4].

**The third step:**

(a) If the "indication" of IE in point 2 is equal to 0 (IE shows *to the left*), this means that the "measurable" point $X$ is on the segment [1, 2]. In this situation, the measurement process ends, because the coordinate of point $X$ ($X \in [1, 2]$) has been determined with the "exactness" equal to the *measurement unit.*

(b) If the "indication" of IE in point 2 is equal to 1 (IE shows *to the right*), this means that the "measurable" point $X$ is on the segment [2, 4]. In this situation, the measurement process continues and IE in the next step enclosing to point 3. As a result, we get two new situations: [2, 3] and [3, 4], depending on the "indication" of IE at the last step of the algorithm.

Figure 1.6 shows the operation of the 3-step *"binary"* algorithm on the segment [0, 8]. The essence of the algorithm is clear from Fig. 1.6 and reduces to enclosing the IE to the middle of the uncertainty interval relative to point $X$ obtained in the previous step on the basis of the "indication" of IE.

Finally, Fig. 1.7 presents an example of the *readout algorithm* widely used in measurement practice. This algorithm underlies the traditional measuring ruler.

The *readout algorithm*, shown in Fig. 1.7, is realized in one step ($n = 1$) and uses three IE, which are enclosed simultaneously to the points of the segment [0, 4], as shown in Fig. 1.7.

### 1.8.4. Restrictions $S$

In the above definition of the ($n$, $k$, $S$)-algorithm, there is an important notion of the *restriction $S$*, which is superimposed on the measurement algorithm. It is clear that any *restriction* reduces the efficiency of the algorithm, but, on the other hand, each *restriction*, as will be shown in the following, leads to very unusual measurement algorithms that may be of theoretical and practical interest.

Let's start from the simplest *restriction* that can be found from a comparative analysis of the classical measurement algorithms, in particular, the *"binary"* *algorithm* (Fig. 1.6) and the *counting algorithm* (Fig. 1.5).

Let's compare the two 3-step $(n, k, S)$-algorithms presented in Figs. 1.5 and 1.6. Both algorithms are implemented in the same number of steps $(n = 3)$ and use only one IE $(k = 1)$.

What is the difference between these $(3, 1, S)$-algorithms? As the comparative analysis shows, the difference between them consists in the character of the movement of IE on the segment $AB$. Indeed, in the example in Fig. 1.6, an arbitrary movement of IE on the segment $[0, 8]$ is allowed, that is, after a particular step, the IE can be enclosing at the next step both *to the right* or *to the left* relative to the point $X$. It is appropriate for the above *restriction* to use the notation $S = 0$. Then, we can consider the so-called $(n, k, 0)$-algorithms, that is, the *measurement algorithms*, in which the movement of IE on the segment $AB$ obeys the restriction $S = 0$.

Let's now consider the *counting algorithm* in Fig. 1.5. In this example, the movement of IE is subjected to the following strict *restriction*: in the process of operation of the algorithm, IE moves on the segment $AB$ only in one direction: from point $A$ to point $B$. In this case, IE participates in the measurement algorithm as long as it is *"to the left"* from point $X$; as soon as IE *"jumps"* over point $X$, it stops its further participation in the measurement algorithm. Denote this *restriction* by $S = 1$. Then we also have the right to consider the so-called $(n, k, 1)$-algorithms, that is, the measurement algorithms, in which the movement of IE on the segment $AB$ obeys the *restriction* $S = 1$.

Note that the *restrictions* $S = 0$ and $S = 1$ are not the only possible *restrictions*. It is important to emphasize that each *restriction* $S$ leads to the discovery of some new class of the *optimal measurement algorithms* having theoretical or practical interest; therefore, the search for reasonable *restrictions*, imposed on the *measurement algorithm*, is an important task of the *algorithmic measurement theory* [16].

It is essential to emphasize once again that the *classical measurement algorithms*, discussed above, played an important role in the development of mathematics and computer science.

In particular, the *counting algorithm* (Fig. 1.5) underlies the *Euclidean definition of natural numbers* (1.10) and the *Eudoxus–Archimedes axiom*; the "*binary*" *algorithm* (Fig. 1.6) underlies the "*binary*" *system* (1.12), without which it is impossible to imagine modern computer science, that is, these *simplest measurement algorithms* affect the foundations of mathematics and computer science.

## 1.9. Optimal $(n, k, 0)$-Algorithms

### 1.9.1. Recursion method

As mentioned above, the $(n, k, 0)$-algorithm is a $n$-step *measurement algorithm*, which uses the $k$ IE, the movement of which on the segment $AB$ that obeys the restriction $S = 0$. Recall that the restriction $S = 0$ essentially means the absence of some restrictions on the movement of IE, that is, each IE at a particular step of the measurement algorithm can be enclosed to the arbitrary point of the segment $AB$.

We will solve the task of synthesis of the "*optimal*" $(n, k, 0)$-algorithm, that is, the measurement algorithm, which, for the given $n$, $k$ and the *restriction* $S = 0$, divides the segment $AB$ into the biggest number of equal intervals.

To solve the problems of synthesizing the optimal $(n, k, S)$-algorithms, we will use the "*method of recurrent relations*" [110], which is widely used to solve combinatorial problems. The essence of this method is that the solution of some $n$-step combinatorial task reduces to solving the $(n - 1)$-step combinatorial task; at the same time, there is the recurrent relation, which connects both solutions. This continues until we come to some "simple" combinatorial task, the solution of which does not cause difficulties.

### 1.9.2. Synthesis of the optimal $(n, k, 0)$-algorithm

The essence of the *method of the recurrent relations* in connection with synthesizing of the *optimal (n, k, 0)-algorithms* is as follows. We do the so-called *inductive hypothesis*. Suppose that for some $n$

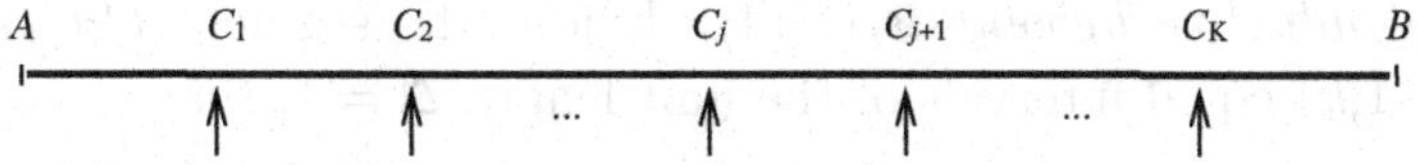

Fig. 1.8. The first step of the optimal $(n, k, 0)$-algorithm.

and $k$, there exists the *optimal* $(n, k, 0)$-*algorithm* that implements on the segment $AB$ the $(n, k)$-*exactness* equal to $F(n, k)$. In other words, our inductive hypothesis consists in the fact that the *optimal* $(n, k, 0)$-*algorithm* divides the segment $AB$ into $F(n, k)$ equal intervals of the unit length $\Delta = 1$, that is,

$$AB = F(n, k). \qquad (1.14)$$

Now, let the first step of the *optimal* $(n, k, 0)$-*algorithm* on the segment $AB$ be in the enclosing of the $k$ *indicatory elements* on the segment $AB$, as shown in Fig. 1.8.

After the first step, by using the *indications* of $k$ IE, we get the $(k + 1)$th situations:

$$(1) \to X \in AC_1;\ (2) \to X \in C_1C_2; \ldots;(j + 1)$$

$$\to X \in C_jC_{j+1}; \ldots; (k + 1) \to X \in C_kB. \qquad (1.15)$$

In this case, all the segments in Fig. 1.8 are connected by the following relationship:

$$AB = AC_1 + C_1C_2 + \cdots + C_jC_{j+1} + \cdots + C_kB. \qquad (1.16)$$

Let's consider the situation $X \in C_jC_{j+1}$. For this situation, before taking the next step, we have $(n - 1)$ steps because one step has already been used, and $k$ IE, which, according to the *restriction* $S = 0$, can be enclosed to some points of the new uncertainty interval $C_jC_{j+1}$, which contains point $X$. But then each of the situations (1.15), that arose after the first step, is not distinguished from the initial situation in Fig. 1.8. Only the number of steps for the realization of the *measurement algorithm*, which operates on the new segment $C_jC_{j+1}$, decreased to $(n-1)$, and then, in this situation, for searching the coordinate of point $X$, we can use the new uncertainty interval $C_jC_{j+1}$ the *optimal* $(n-1, k, 0)$-*algorithm*, which, according to the *inductive hypothesis* (1.14), divides the segment $C_jC_{j+1}$ on $F(n-1,k)$ equal intervals of the unit length $\Delta = 1$, that is,

$$C_jC_{j+1} = F(n-1,k). \qquad (1.17)$$

The above reasoning is valid for all the $(k+1)$ situations (1.15), that is, for some of the segments (1.15), the relation (1.17) is fulfilled. By using now the relation (1.16), connecting the initial segment $AB$ with $(k+1)$ segments (1.15), as well as expressions (1.14) and (1.17), we obtain the following recurrent formula for the $(n, k)$-*exactness* of the *optimal* $(n, k, 0)$-*algorithm*:

$$F(n,k) = (k+1)F(n-1,k). \qquad (1.18)$$

Now, we will try to solve the *recurrent relation* (1.18), that is, to obtain the expression for the *efficiency function* $F(n,k)$ in the explicit form. For this, we use for the function $F(n-1,k)$ in the formula (1.18) the same *recurrent formula* (1.18), that is, we represent the *recurrent formula* (1.18) in the form:

$$F(n,k) = (k+1)(k+1)F(n-2,k). \qquad (1.19)$$

Continuing this process, that is, sequentially decomposing in (1.19) all the expressions for $F(n-2,k)$, $F(n-3,k)$, $\ldots, F(2,k)$ by the recurrent formula (1.18), we get the following expression for (1.18):

$$F(n,k) = (k+1)^{n-1}F(1,k). \qquad (1.20)$$

In the expression (1.20), we have the member $F(1,k)$, whose value is unknown to us. According to our definitions, the member $F(1,k)$ is the $(1,k)$-*exactness* of the *optimal* $(1,k,0)$-*algorithm*, that is, the *measurement algorithm*, which is realized in the first step and uses $k$ IE. It is clear that in this case the *optimal strategy* consists in dividing the last uncertainty interval into $(k+1)$ equal parts by using $k$ IE; the $(1,k)$-*exactness* of this algorithm is equal to

$$F(1,k) = k+1. \qquad (1.21)$$

Substituting the expression (1.21) into (1.20), we obtain the expression for the *efficiency function* of the *optimal* $(n, k, 0)$-*algorithm* in the explicit form:

$$F(n, k) = (k + 1)^n. \tag{1.22}$$

It is easy to understand that the strategy of the *optimal* $(n, k, 0)$-algorithm at each step is very simple: *it is necessary to divide the uncertainty interval by using* $k$ *IE into* $(k + 1)$ *equal parts at each step.*

### 1.9.3. Special cases of the optimal $(n, k, 0)$-algorithm

Let's consider the *particular extreme cases* of the *optimal* $(n, k, 0)$-algorithm. Let $n = 1$. In this case, the formula (1.22) reduces to (1.21), and the *optimal* $(1, k, 0)$-algorithm reduces to the *readout algorithm* (Fig. 1.7).

For the case of $k = 1$, the formula (1.22) reduces to the formula $F(n, 1) = 2^n$, which sets the value of $(n, 1)$-*exactness* of the "*binary*" *algorithm*, i.e., the "*binary*" *algorithm*, considered above (Fig. 1.6), is the *extreme particular case* of the *optimal* $(n, k, 0)$-algorithm.

### 1.9.4. Optimal $(n, k, 0)$-algorithms and positional numeral systems

As shown above, the "*binary*" *algorithm* (Fig. 1.6) "generates" the "*binary*" *system* (1.12). But then it is clear that the *optimal* $(n, k, 0)$-*algorithm* in general case also "generates" some positional numeral system with the base $R = k + 1$, that is, the following representation of the natural number $N$ in the form:

$$N = a_{n-1}R^{n-1} + a_{n-2}R^{n-2} + \cdots + a_i R^i + \cdots + a_1 R^1 + a_0 R^0, \tag{1.23}$$

where $a_i \in \{0, 1, 2, \ldots, k\}$ is the numeral of the $i$th digit.

In particular, for the case of $k = 9$, the formula (1.23) reduces to the classical *decimal* system:

$$A = a_{n-1}10^{n-1} + a_{n-2}10^{n-2} + \cdots + a_i 10^i + \cdots + a_1 10^1 + a_0 10^0.$$

For the case $k = 59$, the formula (1.23) reduces to the *Babylonian positional system with the base of* 60. From these examples and arguments, it follows that the *optimal* $(n, k, 0)$-*algorithms* "generate" all well-known positional numeral systems (including the *Babylonian system with the base* 60, *decimal*, *binary*, and other positional systems).

## 1.10. Optimal $(n, k, 1)$-Algorithms Based on Arithmetic Square

### 1.10.1. Synthesis of the optimal $(n, k, 1)$-algorithm

Now, we synthesize the *optimal* $(n, k, 1)$-*algorithm* [16]. Let's recall that the *restriction* $S = 1$ means that *indicatory elements* (IE) move along segment $AB$ only in one direction: from point $A$ to point $B$. This means that if some IE at a certain step turned out *to the right* of point $X$, then this IE "leaves the game", i.e., it cannot be used further in the measurement process.

The above *"counting" algorithm* (Fig. 1.5), underlying the "Euclidean definition" of the natural number (1.10), is a vivid example of the *restriction* $S = 1$.

Let us carry out the same reasoning as for the synthesis of the *optimal* $(n, k, 0)$-*algorithm*. Suppose that for some $n$ and $k$, there exists the *optimal* $(n, k, 1)$-*algorithm* that realizes on the segment $AB$ the $(n, k)$-*exactness* equal to $F(n, k)$. In other words, the *optimal* $(n, k, 1)$-*algorithm* divides the segment $AB$ into $F(n, k)$ equal intervals of the unit length $\Delta = 1$, that is,

$$AB = F(n, k). \tag{1.24}$$

Now, let the first step of the *optimal* $(n, k, 1)$-*algorithm* on the segment $AB$ be in the enclosing the $k$ *indicatory elements* to the points of the segment $AB$, as shown in Fig. 1.9.

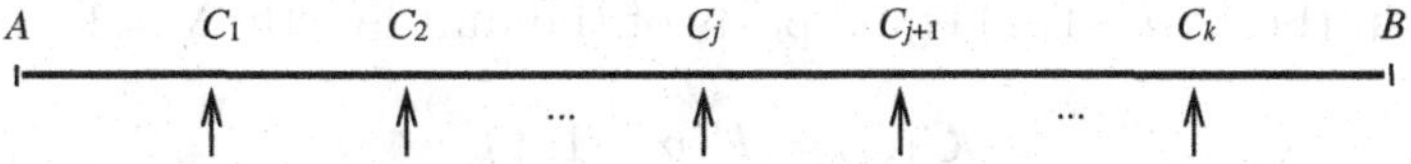

$$A \qquad C_1 \qquad C_2 \qquad\qquad C_j \qquad C_{j+1} \qquad\qquad C_k \qquad B$$

Fig. 1.9. The first step of the optimal $(n, k, 1)$-algorithm.

After the first step, on the basis of the *indications* of $k$ IE, the $(k+1)$th situations may arise:

$$(1) \to X \in AC_1; \; (2) \to X \in C_1C_2; \ldots; (j+1) \to$$
$$X \in C_jC_{j+1}; \ldots; (k+1) \to X \in C_kB, \qquad (1.25)$$

where

$$AB = AC_1 + C_1C_2 + \cdots + C_jC_{j+1} + \cdots + C_kB. \qquad (1.26)$$

Now, let's analyze the situations that may arise after the first step of the $(n, k, 1)$-*algorithm*.

Let's consider the first situation $X \in AC_1$ among the situations (1.25). For this situation, all $k$ IE turned out to be *to the right* of point $X$. This means that, according to the *restriction $S = 1$*, none of the $k$ IE can be further closing to the interval $AC_1$, that is, the measurement actually ends after the first step because all IE turned out *to the right* of point $X$. But, according to our *"inductive hypothesis"*, the *optimal $(n, k, 1)$-algorithm* "divides" the segment $AB$ into $F(n,k)$ equal parts of the unit length $\Delta = 1$. Then, it follows from this reasoning that the first uncertainty interval $AC_1$ must be the segment of the unit length $\Delta = 1$, that is,

$$AC_1 = 1. \qquad (1.27)$$

Let's now consider the second situation $X \in C_1C_2$ among the situations (1.25). In this situation, after the first step, we have the $(n - 1)$ steps and one IE, which is *to the left* of the point $X$; all other IE are *to the right* of point $X$ and, according to the *restriction $S = 1$*, cannot participate in the measurement process. Then we can apply to the segment $C_1C_2$ the *optimal $(n-1, 1, 1)$-algorithm*, which, according to the *"inductive hypothesis"* (1.24), divides the segment

$C_1C_2$ on the $F(n-1,1)$ equal parts of the unit length $\Delta = 1$, that is,

$$C_1C_2 = F(n-1,1). \tag{1.28}$$

Let's now consider the situation $X \in C_jC_{j+1}$ among the situations (1.25). In this situation, we have the $(n-1)$ steps and $j$ IE, which are *to the left* of point $X$ (all other $(k-j)$ IE are *to the right* of point $X$ and, according to the *restriction* $S = 1$, cannot participate in the measurement process). Then we can apply to the segment $C_jC_{j+1}$ the *optimal* $(n-1,j,1)$-*algorithm*, which, according to the "*inductive hypothesis*" (1.24), divides the segment $C_jC_{j+1}$ on the $F(n-1,j)$ equal parts of the unit length $\Delta = 1$, that is,

$$C_jC_{j+1} = F(n-1,j). \tag{1.29}$$

Finally, in the last situation $X \in C_kB$ among the situations (1.25), all the $k$ IE are *to the left* of point $X$; this means that we can enclose to the segment $C_kB$ the *optimal* $(n-1,k,1)$-*algorithm*, which, according to the "*inductive hypothesis*" (1.24), divides the segment $C_kB$ on the $F(n-1,k)$ equal parts of the unit length $\Delta = 1$, that is,

$$C_kB = F(n-1,k). \tag{1.30}$$

Taking into consideration the relation (1.26), as well as the expressions (1.27)–(1.30), we can write the following recurrent relation for the $(n,k)$-*exactness* of the *optimal* $(n,k,1)$-*algorithm*:

$$F(n,k) = 1 + F(n-1,1) + \cdots + F(n-1,j)$$
$$+ \cdots + F(n-1,k-1) + F(n-1,k). \tag{1.31}$$

Now, we consider the sum:

$$1 + F(n-1,1) + \cdots + F(n-1,j) + \cdots + F(n-1,k-1), \tag{1.32}$$

taken from the expression (1.31). According to the recurrent formula (1.31), the sum (1.32) is equal to $F(n,k-1)$, that is,

$$F(n,k-1) = 1 + F(n-1,1) + \cdots + F(n-1,j)$$
$$+ \cdots + F(n-1,k-1). \tag{1.33}$$

By using (1.33), we can simplify the recurrent relation (1.31) and write it in the following compact form:

$$F(n, k) = F(n, k - 1) + F(n - 1, k).  \qquad (1.34)$$

Now let's construct the table of the numbers of $F(n, k)$ by using the recurrent formula (1.34). To do this, we find out the *extreme values* of the function $F(n, k)$, corresponding to the values $n = 0$ and $k = 0$, that is, the values of $F(0, k)$ and $F(n, 0)$. Recall that $F(0, k)$ is the $(0, k)$-*exactness* of the *optimal* $(0, k, 1)$-*algorithm*, and $F(n, 0)$ is $(n, 0)$-*exactness* of the *optimal* $(n, 0, 1)$-*algorithm*. But, according to our definitions, $(0, k, 1)$-algorithm is the $(n, k, 1)$-*algorithm*, in which the number of steps is equal to $n = 0$, and $(n, 0, 1)$-*algorithm* is $(n, k, 1)$-*algorithm*, in which the number of IE is equal to $k = 0$. But, from the "physical" sense of the task, the $(n, k, 1)$-*algorithms*, in which either $n = 0$ or $k = 0$, cannot narrow the initial uncertainty interval, and therefore, for such $(n, k, 1)$-*algorithms*, the $(n, k)$-*exactness* or *efficiency function* is always identically equal to 1, that is,

$$F(0, k) = F(n, 0) = 1.  \qquad (1.35)$$

### 1.10.2.  Arithmetic square

By using the recurrent relation (1.34) and the initial conditions (1.35), we can construct Table 1.1 for the numerical values of the *efficiency function* $F(n, k)$ of the *optimal* $(n, k, 1)$-*algorithm*.

By comparing Table 1.1 with the table for binomial coefficients, known as *arithmetic square* or *Tartaglia rectangle* [132], we come to the unexpected conclusion that the *efficiency function* of $F(n, k)$ is expressed by using the formula for the binomial coefficients:

$$F(n, k) = C_{n+k}^n = C_{n+k}^k.  \qquad (1.36)$$

Niccolò Fontana Tartaglia (1499/1500–1557) (Fig. 1.10) was the famous Italian mathematician and engineer (designing fortifications) [132].

Table 1.1.  Arithmetic square.

|   | 0 | 1 | 2 | 3 | 4 | 5 | $\cdots$ | $n$ |
|---|---|---|---|---|---|---|---|---|
| 0 | 1 | 1 | 1 | 1 | 1 | 1 | $\cdots$ | 1 |
| 1 | 1 | 2 | 3 | 4 | 5 | 6 | $\cdots$ | $F(n,1)$ |
| 2 | 1 | 3 | 6 | 10 | 15 | 21 | $\cdots$ | $F(n,2)$ |
| 3 | 1 | 4 | 10 | 20 | 35 | 56 | $\cdots$ | $F(n,3)$ |
| 4 | 1 | 5 | 15 | 35 | 70 | 126 | $\cdots$ | $F(n,4)$ |
| 5 | 1 | 6 | 21 | 56 | 126 | 252 | $\cdots$ | $F(n,5)$ |
| $\vdots$ | $\vdots$ | $\vdots$ | $\vdots$ | $\vdots$ | $\vdots$ | $\vdots$ | $\vdots$ | $\vdots$ |
| $k$ | 1 | $F(1,k)$ | $F(2,k)$ | $F(3,k)$ | $F(4,k)$ | $F(5,k)$ | $\cdots$ | $F(n,k)$ |

Fig. 1.10.  Italian mathematician and engineer Niccolò Fontana Tartaglia.

Tartaglia published many books, including the first Italian translations of Archimedes and Euclid. He was the first one, who used mathematics for the investigation of the paths of *cannonballs*, known as *ballistics*, in his book *A New Science*.

### 1.10.3.  Optimal $(n, k, 1)$-algorithm

The operation of the *optimal* $(n, k, 1)$-*algorithm* can be demonstrated by using the *arithmetic square* (Table 1.1). Indeed, for the given $n$ and $k$, the value of the *efficiency function* $F(n, k)$ is at the intersection of the $n$th column and the $k$th row of the *arithmetic square* (see Table 1.1).

The first step of the *optimal* $(n, k, 1)$-*algorithm* is as follows. The coordinates of the application points of $k$ IE on the segment $AB$ (the points $C_1, C_2, \ldots, C_j, C_{j+1}, \ldots, C_k$) relative to point $A$ (Fig. 1.8) are in the $n$th column of the *arithmetic square* (Table 1.1), that is,

$$AC_1 = 1, AC_2 = F(n, 1), \ldots, AC_j$$
$$= F(n, j - 1), \ldots, AC_k = F(n, k - 1). \tag{1.37}$$

If after the first step of the $(n, k, 1)$-*algorithm*, the $j$th IE turned to be *to the left* of point $X$, and the remaining $(k - j)$ IE turned to be *to the right* of point $X$, then the *uncertainty interval* relative to point $X$ decreases to the segment $C_j C_{j+1}$. The length of this segment is equal to $F(n - 1, j)$. This binomial coefficient is at the intersection of the $(n-1)$th column and $j$th row of the *arithmetic square* (Table 1.1). To find the binomial coefficient $F(n - 1, j)$, we must move from the initial coefficient $F(n, k)$ in the one column *to the left* (we "lost" one step) and the $(k - j)$ rows up (we "lost" $(k - j)$ IE).

In the second step, we take the point $C_j$ for the new beginning of coordinates. In this case, the second step consists in enclosing the $j$ IE to some points $D_1, D_2, \ldots, D_j$ of the new *uncertainty interval* $C_j C_{j+1}$. The coordinates of the points $D_1, D_2, \ldots, D_j$ relative to the new beginning of coordinates (the point $C_j$) are at the $(n - 1)$th columns of the *arithmetic square* above the binomial coefficient $F(n - 1, j)$, that is,

$$C_j D_1 = 1, \ C_j D_2 = F(n - 1, 1), \ldots, C_j D_3$$
$$= F(n - 1, 2), C_j D_j = F(n - 1, j - 1). \tag{1.38}$$

This process ends when either all the steps of the *measurement algorithm* or all the *indicatory elements* will be *exhausted*.

### 1.10.4. An example of the optimal $(n, k, 1)$-algorithm

As an example, let's consider the operation of the *optimal* $(3, 3, 1)$-*algorithm* on the *initial segment* $[0, 20]$ (Fig. 1.11).

The *optimal* $(3, 3, 1)$-*algorithm* is realized in the three steps and uses three IE in this case. Note that the number $F(3, 3) = 20$, which

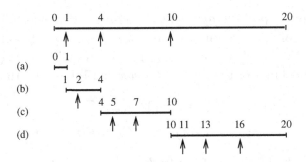

Fig. 1.11. The optimal $(3, 3, 1)$-algorithm.

is located at the intersection of the third column and the third row of the *arithmetic square* (Table 1.1), is equal to the length of the *initial uncertainty interval* for this algorithm (Table 1.1). The coordinates of the applications of IE in the first step of $(3, 3, 1)$-*algorithm* are in the third column of the *arithmetic square* (Table 1.1) above the number 20 (in bold).

**The first step** of the $(3, 3, 1)$-*algorithm* on the *initial segment* $[0, 20]$ consists in the enclosing of the three IE to points 1, 4, 10 (Fig. 1.10). After the first step, based on the *indications* of the IE, four situations may arise (Figs. 1.10(a)–(d)).

**The second step**: Situation (a): For this situation, the measurement process ends because all the IE are *to the right* of point $X$.

Situation (b): For this situation, we have only one IE, which is enclosed to point 2.

Situation (c): For this situation, we have two IE, which are enclosed to points 5 and 7, respectively.

Situation (d): For this situation, we have three IE. In this case, point 10 is the new beginning of the coordinates and three IE are enclosed to points 1, 3, 6 relative to the new beginning of coordinates (point 10). By summing up the numbers 1, 3, 6 to the number 10, we get the coordinates of the enclosed points of three IE at the next step: 11, 13, 16.

After the second step, the following situations may arise: $[1, 2]$, $[2, 4]$, $[4, 5]$, $[5, 7]$, $[7, 10]$, $[10, 11]$, $[11, 13]$, $[13, 16]$, $[16, 20]$. Note that for situations [1, 2], [4, 5], [10, 11], the measurement process ends at the second step.

**The third step:** For the situations [2, 4], [5, 7] and [11, 13], the third step is the enclosing of the one IE to points 3, 6, 12, respectively.

For the situations [7, 10] and [13, 16], the third step consists in the enclosing of the two IE to points 8, 9 or 14, 15, respectively.

For the situation [16, 20], the third step consists in the enclosing of the three IE to points 17, 18, 19.

### 1.10.5. The extreme particular cases of the optimal $(n, k, 1)$-algorithm

Now, we will consider the *extreme special cases* of the *optimal* $(n, k, 1)$-*algorithm*, that is, when $n = 1$ or $k = 1$. For the case $n = 1$, the expression (1.36) takes the following form:

$$F(1, k) = C_{1+k}^1 = C_{1+k}^k = k + 1. \tag{1.39}$$

It is easy to verify that in the case of $n = 1$, the *optimal* $(1, k, 1)$-*algorithm* reduces to the *readout algorithm* discussed above (Fig. 1.7).

For the case $k = 1$, the expression (1.36) takes the following form:

$$F(n, 1) = C_{n+1}^n = C_{n+1}^1 = n + 1. \tag{1.40}$$

It is easy to verify that in the case of $k = 1$, the *optimal* $(n, 1, 1)$-*algorithm* reduces to the *counting algorithm* discussed above (Fig. 1.5).

### 1.10.6. The importance of the binomial algorithms for mathematics and computer science

Thus, the main mathematical result, obtained in Ref. [16] at the synthesis of the *optimal* $(n, k, 1)$-*algorithm*, is the fact that these studies led us to the *arithmetic square* and *binomial coefficients*. We emphasize that this result is much unexpected because by starting the synthesis of this algorithm, we did not assume its connection with the *binomial coefficients*. This gives us a reason to call the $(n, k, 1)$-algorithms the *binomial algorithms of measurement*.

What is the significance of the *binomial measurement algorithms* for mathematics? Here, it is important to emphasize that the *optimal* $(n, k, 1)$-*algorithm* is a generalization of the *counting algorithm* (Fig. 1.5), which, as mentioned above, historically underlies the *elementary number theory*, the fundamentals of which are outlined in Euclid's *Elements*. Therefore, we can assume that the above-considered *binomial measurement algorithms* can be used for further development of the *elementary number theory* created by the ancient Greeks.

There is another idea arising from the interpretation of the *optimal* $(n, k, S)$-*algorithms* as *new positional representations of natural numbers*. With this approach, the *binomial algorithms* can be of practical importance for modern computer science. And maybe some of our readers, who can be fascinated by such unusual numeral system, will design a new (*binomial*) computer. Those readers, who are interested in the *binomial computers* can refer to the book [136] of the Ukrainian scientist Alexey Borisenko (Sumy University).

# Chapter 2

# Principle of Asymmetry of Measurement and Fibonacci Algorithms of Measurement

## 2.1. Bachet–Mendeleev Problem

The constructive approach, at once, raises questions about the *optimization problems* in the measurement theory, which were not a topical task in the *classical measurement theory*. By studying the *optimization problems in the measurement theory*, we unexpectedly come to the combinatorial task, known as the *task about the choice of the best weight systems* (the "*task of optimal weights*"). For the first time, this task was described in Fibonacci's 1202 book *Liber abaci*. From Fibonacci's work, this task was moved to Luca Pacioli's 1494 book *Summary of Arithmetics, Geometry, Proportion and Proportionality*. After Pacioli's work, the "*task of optimal weights*" again appeared in the 1612 book *Collection of the Pleasant and Entertaining Problems* by the French mathematician *Claude Gaspard Bachet de Méziriac* (1581–1638).

Claude Gaspard Bachet (Fig. 2.1) was born in 1581 in Bourg-en-Bresse in Savoy (France). This was a region, which belonged to France, Spain or Italy in different periods. Therefore, the life of Claude Gaspard Bachet is connected with France, Spain and Italy of that period. He was a writer of books on *mathematical puzzles* and *tricks* that became the basis for all the books on *mathematical puzzles*. Bachet's 1612 book *Collection of the Pleasant*

Fig. 2.1.  Claude Gaspard Bachet de Méziriac (1581–1638).

*and Entertaining Problems* contains various mathematical problems. Fibonacci's *"task of optimal weights"* was one of them. This task was formulated as follows: *"What is the least number of weights used to weigh any whole number of pounds from 1 to 40 inclusively by using a balance if the standard weights can be placed on both cups of balances?"*

In the Russian historical–mathematical literature, Fibonacci's *"task of optimal weights"* is also known by the name of the *Bachet–Mendeleev problem* in honor of Bachet de Méziriac and the famous Russian chemist, Dmitry Mendeleev (Fig. 2.2).

However, this raises the following question: Why did the great Russian chemist Mendeleev show interest in Fibonacci's *"task of optimal weights"*? The answer to this question follows from some little-known facts from the scientific life of the great scientist. In 1892, Mendeleev was appointed as the Director of the Russian Depot of Standard Weights, which, based on his own initiative, was transformed in 1893 to the Main Board of Weights and Measures of Russia.

Mendeleev remained its Director until the end of his life. Thus, the final stage of Mendeleev's life (since 1892 until his death in 1907) was connected deeply with the development of metrology. During this period, Mendeleev was actively involved in the various problems, connected with *measures, measurement,* and *metrology;* the *"task of*

Fig. 2.2. Dmitry Mendeleev (1834–1907).

*optimal weights"* was one of them. Mendeleev's contribution to the development of metrology in Russia was so great that he was named *"the father of Russian metrology"* and the *"task of optimal weights"* was named the *Bachet–Mendeleev problem*.

The essence of the problem consists of the following [16]. Suppose, we need to weigh the balances of some integer-valued weight $Q$ in the range from 0 up to $Q_{\max}$ by using the $n$ standard weights $\{q_1, q_2, \ldots, q_n\}$, where $q_1 = 1$ is a measurement unit, $q_i = k_i \times q_1$, and $k_i$ is some natural number. It is clear that the maximal weight $Q_{\max}$ is equal to the sum of all the standard weights, that is,

$$Q_{\max} = q_1 + q_2 + \cdots + q_n = (k_1 + k_2 + \cdots + k_n)q_1. \qquad (2.1)$$

We have to find the *Optimal System of Standard Weights*, that is, such standard weight system, which gives the *maximal value of the sum* (2.1) for the given measurement unit $q_1 = 1$ among all possible variants.

There are two variants of the solution of the *Bachet–Mendeleev problem*. In the first case, the standard weights can be placed only on the free cup of balances; in the second case, they can be placed on both cups of balances. The *Binary Measurement Algorithm* with the binary system of standard weights $\{1, 2, 4, 8, \ldots, 2^{n-1}\}$ is the *optimal solution* for the first case. It is clear that the *binary measurement algorithm* generates the *binary method of number representation*:

$$Q = \sum_{i=0}^{n-1} a_i 2^i, \tag{2.2}$$

where $a_i \in \{0, 1\}$ is a binary numeral.

Note that the binary numerals 0 or 1 have a precise measurement interpretation. Note that the binary numeral 0 or 1 encodes one of the two possible positions of the balances; thus, $a_i = 1$ if the balances remain in the initial position after application of the next standard weight on the free cup, otherwise $a_i = 0$.

The ternary system of the standard weights $\{1, 3, 9, 27, \ldots, 3^{n-1}\}$ is the optimal solution to the second variant of the "weighing" problem, when we can put the standard weights on both cups of the balances. It is clear that the *ternary system of the standard weights* results in the *ternary method of number representation*:

$$Q = \sum_{i=0}^{n-1} b_i 3^i, \tag{2.3}$$

where $b_i$ is a ternary numeral that takes the values from the set $\{-1, 0, 1\}$.

Hence, by solving the *"task of optimal weights"*, Fibonacci and his followers found a deep connection between algorithms of measurement and positional methods of number representation. This is the key idea of the *algorithmic measurement theory* [16], which goes back in its origin to the Fibonacci's *"task of optimal weights"*. Thus, from the measurement problem, we unexpectedly come to the positional method of number representation, the greatest mathematical discovery of Babylonian mathematics.

## 2.2. Asymmetry Principle of Measurement

The principles of *finiteness* and *potential feasibility* of measurement, underlying the constructive (algorithmic) measurement theory, are "external" with respect to the measurement. They have such a general character that there is a danger of reduction of the measurement to some trivial result (for example, to the above *binary algorithm* of measurement), which reduces to the consecutive comparison of the measurable weight $Q$ with the binary standard weights: $2^{n-1}$, $2^{n-2}, \ldots, 2^0$, by starting from the higher standard weight $2^{n-1}$.

For obtaining non-trivial results, the methodological basis of the *constructive (algorithmic) theory of measurement* should be added by a certain general principle, which follows from the essence of measurement itself. Such a principle follows directly from the analysis of the *binary algorithm of measurement*, which is the *optimal solution* of the *Bachet–Mendeleev problem*.

We will analyze the above *binary algorithm of measurement* with the example of weighing the measurable weight $Q$ by using standard weights $2^{n-1}$, $2^{n-2}, \ldots, 2^0$ (Fig. 2.3).

This analysis allows finding a measurement property of general character for any thinkable measurement based on the comparison of the measurable weight $Q$ with the standard weights.

We will examine the weighing process of the unknown weight $Q$ on the balances by using the *binary standard weights* $2^{n-1}$, $2^{n-2}, \ldots, 2^0$. In the first step of the *binary algorithm*, the higher standard weight $2^{n-1}$ is placed on the free cup of balances (Fig. 2.3(a)). The balances compare the weight $Q$ with the higher standard weight $2^{n-1}$. It is clear that after the first step, we can get two possible situations: $2^{n-1} < Q$ (Fig. 2.3(a)) and $2^{n-1} \geq Q$ (Fig. 2.3(b)). In the first situation (Fig. 2.3(a)), the second step is to place the next standard weight $2^{n-2}$ on the free cup of balance.

In the second situation (Fig. 2.3(b)), the weigher should perform two operations. In the situation in Fig. 2.3(b), we should remove the standard weight $2^{n-1}$ from the free cup of balances, as shown in Fig. 2.3(b); after that, the balance should return to the initial position, as shown in Fig. 2.3(c). After returning the balances to the initial position, the next standard weight $2^{n-2}$ should be placed on the free cup of balances, as shown in Fig. 2.3(c).

As we can see from Fig. 2.3, for both the situations in Fig. 2.3(a) and Fig. 2.3(b), the weigher's actions on the next algorithm step are distinguished in their complexity. In the case of Fig. 2.3(a), the weigher fulfills only one operation, that is, he adds the second

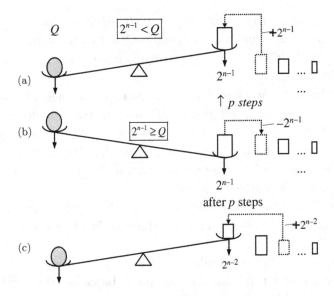

Fig. 2.3. Asymmetry principle of measurement.

standard weight $2^{n-2}$ to the free cup of balances. In the second case (Fig. 2.3(b)), the weigher's actions are determined by two factors. First, he should remove the first standard weight $2^{n-1}$ from the free cup of balances, and after that, the balances are returned back to their initial position. After the balances return to their initial position, the weigher has to place the next standard weight $2^{n-2}$ on the free cup of balances. Thus, in the first case, the weigher has to perform only one operation, that is, to place the standard weight $2^{n-2}$ on the free cup of balances in the next step. However, in the second case (Fig. 2.3(b)), the weigher has to perform two sequential operations:

(1) remove the previous standard weight $2^{n-1}$ from the free cup, as shown in Fig. 2.3(b);
(2) place a new standard weight $2^{n-2}$ on the free cup of balances, as shown in Fig. 2.3(c).

In the second case (Fig. 2.3(b) and 2.3(c)), the weigher's actions are more complicated in comparison with the first case (Fig. 2.3(a)) because the weigher has to remove the previous standard weight $2^{n-1}$ and takes into consideration the time necessary for the balances to return to their initial position. By synthesizing the optimal measurement algorithms, we have to take into consideration these new data, which influence the process of measurement. We name this discovered property of measurement the *Asymmetry Principle of Measurement* [16].

## 2.3. A New Formulation of Bachet–Mendeleev Problem

Let's introduce the above measurement property (*"asymmetry principle of measurement"*) to the *"task about the optimal weighing"* proposed by Fibonacci. For this purpose, we will consider the measurement as a process, occurring at a discrete time; and let the operation of *"adding standard weight"* be performed in one unit of discrete time and the operation of *"removing standard weight"* (which is accompanied by the return of the *lever balances* to their initial position) be performed in the $p$ units of discrete time, where $p \in \{0, 1, 2, 3, \ldots\}$.

It is clear that the parameter $p$ models the *"inertia"* of the *lever balances*. The case $p = 0$ corresponds to the "ideal situation", when we neglect the *"inertia"* of the *lever balances*. It was this case that Fibonacci and all his followers, including Pacioli, Bachet and Mendeleev, studied. For the remaining cases $p > 0$, we get the new versions of the *"task about optimal weighing"*, the solution of which is the main goal of the *algorithmic measurement theory* [16].

A further generalization of the *Bachet–Mendeleev task* is to increase the number of *"lever balances"* from 1 to $k$ ($k$ is a positive integer), where the same weight $Q$ is placed on the left-hand bowls of all balances. This situation corresponds to the case of *"parallel measurements"*, when one and the same measurable weight $Q$ is compared with *"standard weights"* using *"comparators"* (this technique is widely used for measuring the electrical magnitudes).

In this case, the generalized *"weighing problem"* can be formulated as follows: it is required to find the *"optimal"* $n$-step *weighing* or *measuring algorithm* by using a system of $k$ balances (*"comparators"*), having the *"inertia"* of $p$ for the condition, when at each step of measurement, the *standard weights* are allowed to put on those and only those *"free cups"*, which are in the initial position *"more"* on this step.

In this case, the generalized *"weighing problem"* can be formulated as follows: it is required to find the *"optimal"* step-by-step weighing algorithm by using a system of the $k$ *"lever balances"* (*"comparators"*), which have the *"inertia"* of $p$, for the condition, when at each measurement step, the next *standard weights* are allowed to be placed on the free cups of those and only those *"lever balances"* (*"comparators"*), which are in the initial position at this step.

It is clear that such task is incomparably significantly more complex than the classical *"weighing problem"* (formulated by Fibonacci in the early 13th century), which is a particular case ($k = 1$ and $p = 0$) of the above generalized, formulated *"weighing problem"*.

## 2.4. Synthesis of the Optimal Fibonacci's Algorithm of Measurement

### 2.4.1. Fibonacci measurement algorithms

Now, let us proceed to discuss the most unexpected result of the *algorithmic measurement theory* [6, 16], namely, the *Fibonacci measurement algorithms*.

Let's turn again to the mathematical model of measurement based on the *"indicatory elements"* and try to introduce a restriction on the measurement, following the above formulated *principle of measurement asymmetry* (Fig. 2.4).

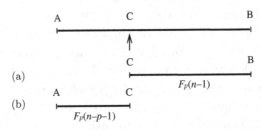

Fig. 2.4. The first step of the Fibonacci measurement algorithm.

This principle implies the following "restriction" of the movement of IE along the segment $AB$. Let IE be enclosed to point $C$ in the first step of the $(n, k, S)$-*algorithm* (Fig. 2.4). Then two situations (a) and (b) could arise, as shown in Fig. 2.4. It is clear that in the situation in Fig. 2.4(a), we can enclose IE to some point on the segment $CB$ at the next step. However, for the situation in Fig. 2.4(b), according to the *principle of measurement asymmetry*, we cannot enclose IE to the points on the segment of $AC$ because the *lever balances* must return to their initial position within $p$ discrete units of the time (see Fig. 2.3). Thus, the "*restriction*" on the *measurement algorithm*, arising from the *principle of measurement asymmetry*, consists in the fact that for the situation in Fig. 2.4(b), it is *forbidden* to enclose IE to points of the segment $AC$ during the $p$ subsequent steps of the algorithm.

Now, we use the same technique that we used successfully for the synthesis of the *optimal* $(n, k, 0)$- and $(n, k, 1)$-*algorithms*. Suppose that for the given $p$ for some $n$, there exists the *optimal n-step algorithm*, which realizes on the segment $AB$ the $(n, 1)$-*exactness*, equal to $F_p(n)$; in other words, our inductive hypothesis consists in the fact that the *optimal n-step measurement algorithm* divides the segment $AB$ into $F_p(n)$ equal intervals of the unit length $\Delta = 1$, that is,

$$AB = F_p(n). \tag{2.4}$$

Suppose now that the first step of the *optimal n-step algorithm*, that satisfies the inductive hypothesis (2.4), consists in closing IE to some point $C$ (Fig. 2.4). Then, after the first step, two situations may arise: (a) $X \in AC$; (b) $X \in CB$.

Let's start with the situation (b): $X \in CB$. On the language of the "*lever balances*", the situation (b) on Fig. 2.4 means that the balances remain in the initial position after the first step and then, according to the *principle of measurement asymmetry*, we can enclose IE to points of the segment $AC$ on the next step. In this case, acting on the segment of $CB$ by the *optimal* $(n-1)$-*step measurement algorithm*, according to the inductive hypothesis (2.4), we can divide the segment $CB$ on the $F_p(n-1)$ equal segments of the unit length

$\Delta = 1$. On the language of "*lever balances*", this situation means that after the first step, the "*lever balances*" remain in the initial position and therefore, according to the *principle of measurement asymmetry*, in the situation in Fig. 2.4(a), we can enclose IE to points of the segment $CB$ on the next step.

In this case, acting on the segment of $CB$ by the *optimal* $(n-1)$-*step measurement algorithm*, according to the inductive hypothesis (2.4), we can divide the segment $CB$ on the $F_p(n-1)$ equal segments of the unit length $\Delta = 1$, that is,

$$CB = F_p(n-1). \tag{2.5}$$

Let's now consider the situation $X \in AC$ (Fig. 2.4). In the language of "*lever balances*" (Fig. 2.3), this situation means that the *balances* went into the opposite state (Fig. 2.3(b)) and therefore, according to the *principle of measurement asymmetry*, we must skip the $p$ steps until the "*lever balances*" return to their initial position. In the language of the model of "*indicatory elements*" (Fig. 2.4), this means that the "*indicatory elements*" could be enclosed to points of the segment $AC$ only after $p$ steps. Because we now have only $(n-1)$ steps, in this situation, our subsequent actions depend on the relationship between the numbers $p$ and $(n-1)$.

Let's assume that $p \geq (n-1)$. In this case, all our steps will be "*exhausted*" before the "*lever balances*" move to their initial position, that is, the measurement process actually ends there. Then, from "physical considerations", it follows that the segment $AC$ must be the segment of the unit length $\Delta = 1$, that is, for the case $p \geq (n-1)$, we will have

$$AC = 1. \tag{2.6}$$

Let's now consider the case $p < (n-1)$. Then in the course of $p$ steps of measurement after the first step, the "*lever balances*" would return to the initial position, that is, after $(p+1)$ steps from the beginning of the measurement, according to the "*principle of measurement asymmetry*", we can continue the measurement process by means of enclosing the "*indicatory elements*" to the points of the segment $AC$ (Fig. 2.4(b)). But, in this case, we only have $(n - p - 1)$ steps to realize the *measurement algorithm*. By acting the *optimal* $(n-p-1)$-*step algorithm*, according to the same inductive hypothesis (2.4), we can divide the segment $AC$ on the $F(n - p - 1)$ equal parts of the unit length $\Delta = 1$, that is,

$$AB = F(n - p - 1). \tag{2.7}$$

Since $AB = AC + CB$, then, by using the results (2.4)–(2.7), we get the following recurrent relation for the calculation of the *n-exactness* of the *optimal n-step measurement algorithm*:

$$F_p(n) = \begin{cases} n + 1 & \text{for } p \geq n - 1, \\ F_p(n - 1) + F_p(n - p - 1) & \text{for } p < n - 1. \end{cases} \tag{2.8}$$

### 2.4.2. Special cases

Let's consider the special cases of the formula (2.8). Let $p = 0$. Then, for this case, the formula (2.8) reduces to the following recurrent relation:

$$F_0(n) = 2F_0(n - 1), \tag{2.9}$$

$$F_0(1) = 2. \tag{2.10}$$

It is clear that the recurrent formula (2.9) for the initial condition (2.10) "generates" the following binary sequence: $2, 4, 8, \ldots, F_0(n) = 2^{n-1}$. It is also clear that the measurement algorithm, corresponding to this case, reduces to the classical "binary" algorithm studied in Chapter 1 (Fig. 1.4).

Let $p = \infty$. For the task, formulated above, this means that IE is "out of the game" as soon as it will be *to the right* of point $X$. For this case, the formula (2.8) takes the form: $F_p(n) = n + 1$, and the above measurement algorithm coincides with the classical *counting algorithm* studied in Chapter 1 (Fig. 1.4).

Table 2.1.  Efficiency function $F_p(n)$.

| $F_p(n)/n$ | 1 | 2 | 3 | 4 | 5 | 6 | 7 | 8 | 9 |
|---|---|---|---|---|---|---|---|---|---|
| $p = 0 : F_0(n)$ | 2 | 4 | 8 | 16 | 32 | 64 | 128 | 256 | 512 |
| $p = 1 : F_1(n)$ | 2 | 3 | 5 | 8 | 13 | 21 | 34 | 55 | 89 |
| $p = 2 : F_2(n)$ | 2 | 3 | 4 | 6 | 9 | 13 | 19 | 28 | 41 |
| $p = 3 : F_3(n)$ | 2 | 3 | 4 | 5 | 7 | 10 | 14 | 19 | 26 |
| $\vdots$ | $\vdots$ | $\vdots$ | $\vdots$ | $\vdots$ | $\vdots$ | $\vdots$ | $\vdots$ | $\vdots$ | $\vdots$ |
| $p = \infty : F_{p=\infty}(n)$ | 2 | 3 | 4 | 5 | 6 | 7 | 8 | 9 | 10 |

Let's consider the case: $p = 1$. For this case, the recurrent formula (2.8) takes the following form:

$$F_1(n) = F_1(n-1) + F_1(n-2) \quad \text{for } n > 2, \qquad (2.11)$$

$$F_1(1) = 2; \quad F_1(2) = 3. \qquad (2.12)$$

If we now calculate the *"efficiency function"* $F_1(n)$ according to (2.11) and (2.12), then we get the following numerical sequence: $2, 3, 5, 8, 13, 21, 34, \ldots$, which is the *Fibonacci sequence*, discovered by Fibonacci when solving the *task of rabbits reproduction*! This fact was the reason why the *optimal measurement algorithm*, defined by the recurrent formula (2.8), was called the *Fibonacci measurement algorithm* [16].

In Table 2.1, the values of the *"efficiency function"* $F_p(n)$ are given for the *optimal Fibonacci algorithms* corresponding to the different values of $p$.

Note that the numerical sequences, given in Table 2.1, starting with $p \geq 2$, represent a new class of numerical sequences, in which each member $F_p(k)$, starting with $k \geq p + 1$, is equal to the sum of the two previous members: $F_p(k - 1)$ and $F_p(k - p - 1)$.

### 2.4.3.  The system of the standard weights for Fibonacci measurement algorithms

Now, let's go to the system of standard weights for the *Fibonacci measurement algorithms*. It has been proved [16] that for this case, the system of *"optimal weights"* $W_p(n)$ is given by the following

Table 2.2. Systems of the "optimal weights" for the Fibonacci algorithms.

| $W_p(n)/n$ | 1 | 2 | 3 | 4 | 5 | 6 | 7 | 8 | 9 |
|---|---|---|---|---|---|---|---|---|---|
| $p = 0 : W_0(n)$ | 1 | 2 | 4 | 8 | 16 | 32 | 64 | 128 | 256 |
| $p = 1 : W_1(n)$ | 1 | 1 | 2 | 3 | 5 | 21 | 34 | 55 | 89 |
| $p = 2 : W_2(n)$ | 1 | 1 | 1 | 2 | 3 | 4 | 6 | 9 | 13 |
| $p = 3 : W_3(n)$ | 1 | 1 | 1 | 1 | 2 | 3 | 4 | 5 | 7 |
| $\vdots$ | $\vdots$ | $\vdots$ | $\vdots$ | $\vdots$ | $\vdots$ | $\vdots$ | $\vdots$ | $\vdots$ | $\vdots$ |
| $p = \infty : W_{p=\infty}(n)$ | 1 | 1 | 1 | 1 | 1 | 1 | 1 | 1 | 1 |

formula:

$$W_p(n) = W_p(n-1) + W_p(n-p-1) \quad \text{for } n > p+1, \quad (2.13)$$

$$W_p(1) = W_p(2) = \cdots = W_p(p+1) = 1. \quad (2.14)$$

The analysis of the expressions (2.13) and (2.14) and Table 2.1 leads us to an unexpected conclusion. It turns out that the recurrent relation (2.13) under the initial conditions (2.14) defines the numerical sequences, obtained in Chapter 3 of Volume I from the study of the so-called *diagonal sums* of Pascal's triangle, named in [16] as *Fibonacci p-numbers*!

Table 2.2 defines the different systems of "*optimal weights*" $W_p(n)$ corresponding to the different values of $p = 0, 1, 2, 3, \ldots, \infty$.

## 2.5. Example of the Fibonacci Algorithm for the Case $p = 1$

Let's now consider the example of the *n-step "optimal" Fibonacci algorithm*, for which the system of "optimal weights" is given by expressions (2.13) and (2.14). Let $p = 1$ and $n = 5$. Let's consider the 5-*step "optimal" Fibonacci measurement algorithm* (Fig. 2.5) corresponding to this case.

From Table 2.1, it follows that the 5-*step Fibonacci algorithm* divides the initial segment $[0, 13]$ into 13 equal parts. To realize this algorithm, we need the five *standard weights*, which according to Table 2.2 have the following values: $\{1, 1, 2, 3, 5\}$.

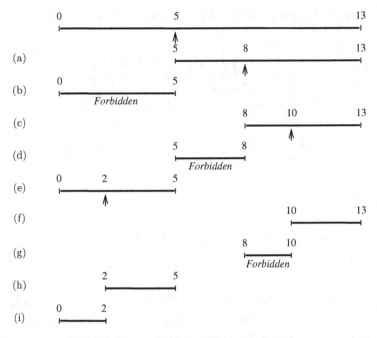

Fig. 2.5. Example of the Fibonacci algorithm.

Let's consider the first three steps of the Fibonacci algorithm, as shown in Fig. 2.5.

**The first step** of the *Fibonacci 3-step algorithm* consists in the enclosing of the *"indicatory element"* to point 5 of the segment [0, 13] (Fig. 2.5). It can be seen that the first step consists in the division of the segment [0, 13] in the Fibonacci relation: $13 = 5 + 8$. After the first step, two situations arise: (a) $(X \in [5, 13])$ and (b) $(X \in [0, 5])$.

**The second step:** (a) For the situation (a) $(X \in [5, 13])$, we choose the next *standard weight* of 3 and, with the help of the *"indicatory element"*, divide the segment [5, 13] of the length of $8 = 13 - 5$ in the Fibonacci relation: $8 = 3 + 5$.

(b) After the second step in situation (b), two new situations arise:

(c) $(X \in [8, 13])$ and (d) $(X \in [5, 8])$. For situation (b) $(X \in [0, 5])$, the second step is *"empty"* because according to the *"restriction" S for the Fibonacci algorithm*, the enclosing of the "*indicative element*" to points of the segment [0, 5] at the second step is *forbidden*.

**The third step:** (c) For situation (c), we use the next standard weight of 2 (see Table 2.2) and divide the segment [8, 13], having the length $5 = 13 - 8$, by using the "*indicative element*" in the Fibonacci relation: $5 = 2 + 3$. Then after the third step, two new situations arise: (e) and (f).

(e) For situation (e), we return to the situation (b), which arose after the second step. In accordance with the "*restriction*" *S for the Fibonacci algorithm*, we can enclose the "*indicative element*" in the third step to the segment [0, 5]. In this case, we can use the standard weight of 2 and divide the segment [0, 5] of the length 5 by using the "*indicative element*" in the Fibonacci relation: $5 = 2 + 3$. Then two new situations, (h) and (i), arise after the third step. It is very easy to trace the operation of the *Fibonacci algorithm* in the next two steps.

Thus, from this example, it can be seen that the essence of the *Fibonacci measurement algorithm* consists in the sequential division of the "*uncertainty interval*", obtained in the previous steps, in the *Fibonacci relations*. It is very simple to prove that this is a *general principle*, which is valid for every value of $p$. In this case, the division of the "*uncertainty interval*" is carried out according to the *recurrent relation* for the *Fibonacci p-numbers*.

## 2.6. The Main Result of the Algorithmic Measurement Theory

As mentioned above, further generalization of the Bachet–Mendeleev problem, based on the "*principle of measurement asymmetry*" (Fig. 2.1), is as follows. Let's increase the number of *lever balances*, involved in the measurement, from 1 to $k$. Let one and the same magnitude $Q$ be *on the left* side of all *lever balances*. This corresponds to the case of *parallel measurement* of the magnitude $Q$ by using $k$ *lever balances* in $n$ steps. The case of "*parallel*" *measurements* is used very often in analog-to-digital converters of electrical magnitudes to increase their speed. In the latter case, the measurable electrical magnitude $Q$ enters parallel to the inputs of all the "comparators" (the electrical analog of "*lever balances*").

Let all the *lever balances*, involved in the measurement, have the *inertia* of $p$ ($p = 0, 1, 2, 3, \ldots$). Let's now synthesize the *optimal* $(n, k, S)$-*algorithm* of measurement, which uses the $k$ *lever balances* (*comparators*), which have the *inertia* $p$.

To simulate the *inertia* of *lever balances* (which correspond to the *indicatory elements* in our model of measurement in Fig. 1.8), we introduce the concept of the *state* of the $j$th *indicatory element* at the $l$th step ($j = 1, 2, 3, \ldots k$; $l = 1, 2.3, \ldots, n$). We denote this *state* by $p_j(l)$. The concept of the *state* has the following *physical* interpretation. If at the $l$th step, the $j$th *lever balances* (*comparator*), corresponding to the $j$th *indicatory element*, are in the initial state, *more* (this means that the $j$th *indicatory element* is *to the left* of point $X$), then it is believed that at the $l$th step, the $j$th *lever balances* (*comparator*) are in the state $p_j(l) = 0$; if at the $l$th step, the $j$th *lever balances* (*comparator*) passed to the opposite state, *less* (that is, the $j$th *indicatory element* is *to the right* of point $X$), then it is believed that $p_j(l) = p$, where $p$ is a given integer, which characterizes the *inertia* of *lever balances* (*comparator*).

If $p_j(l) > 0$, this means that $j$th *lever balances* are in the stage of passage to the initial state and, therefore, the $j$th IE cannot be enclosed to points of the segment $AB$. By starting from the $(l+1)$th step, the *state* of the $j$th IE with each step decreases by 1, that is, $p_j(l + 1) = p - 1$.

In the general case, at the first step of the algorithm, *indicatory elements* can be in some *initial states*:

$$p_1, p_2, \ldots, p_t, p_{t+1}, \ldots, p_{k-1}, p_k. \tag{2.15}$$

Recall that the *initial states* (2.15) at each step can take the values from the set

$$p_t \in \{0, 1, 2, 3, \ldots, p\}; \quad t = 1, 2, 3, \ldots, k. \tag{2.16}$$

For this case, before each step, the renumbering of the *indicatory elements* occurs in such a way that their *initial states* (2.15) would be in the following relationships:

$$p_1 \leq p_2 \leq \cdots \leq p_t \leq p_{t+1} \leq \cdots \leq p_{k-1} \leq p_k. \qquad (2.17)$$

Later on, the $(n, k, S)$-algorithm, in which the *initial states* of the *indicatory elements* (2.15) satisfy the relationships (2.17), will be named $(n; p_1, p_2, \ldots, p_t, p_{t+1}, \ldots, p_{k-1}, p_k)$-algorithm.

Let's denote the *efficiency function* of the *optimal* $(n; p_1, p_2, \ldots, p_t, p_{t+1}, \ldots, p_{k-1}, p_k)$-*algorithm* as follows:

$$F_p(n, k) = F_p(n; p_1, p_2, \ldots, p_t, p_{t+1}, \ldots, p_{k-1}, p_k). \qquad (2.18)$$

Note that at each step, only those *indicatory elements* that are in state 0 can be enclosed to points of the segment $AB$; the remaining *indicatory elements* reduce their *states* by 1.

Let at the first step of the *optimal* $(n, k, S)$-*algorithm*, the *initial states* of the *indicatory elements* be the following:

$$p_1 = p_2 = \cdots = p_t = 0; \quad p_{t+1} > 0. \qquad (2.19)$$

Then, we can denote the *efficiency function* (2.18) of such $(n, k, S)$-*algorithm* as follows:

$$F_p(n, k) = F_p \left( n; \underbrace{0, 0, \ldots, 0}_{t}, p_{t+1}, \ldots, p_{k-1}, p_k \right). \qquad (2.20)$$

From the definition (2.20), it follows that the *efficiency function* depends not only on the number of steps $n$ and the number of *indicatory elements* $k$ of the $(n, k, S)$-*algorithm* but also on the *initial states* of the *indicatory elements*, given by (2.19).

In the synthesis of the *optimal* $(n, k, S)$-*algorithm*, we must take into consideration the following circumstances. First of all, in the first step, only those *indicatory elements* from (2.19) which are in the *state* of 0 (that is, the *indicatory elements* from the first to the tth one) can be enclosed to the segment $AB$. In addition, after the first step, all the *indicatory elements*, which proved to be *to the left* of point $X$ in Fig. 2.4, remain in the *state* 0, and all IE, which proved to be *to the right* of point $X$, pass to the *state p*. Besides, when the first step is completed, the *initial states* (2.19) of all other IE (from $(t+1)$th to $k$th IE) decrease by 1.

Now, if we put in order the *indicatory elements* before the second step according to (2.17), then we will get the following set of *initial states* of the *indicatory elements* at the second step:

$$\underbrace{0,0,\ldots,0}_{r},p_{t+1}-1,\ldots,p_k-1,\underbrace{p,p,\ldots,p}_{t-r}.$$

We will not delve deep into the solution of this mathematical task. Its solution is set forth in the Russian language book by Alexey Stakhov [16] and the English language book by the same author [6]. In Refs. [6, 16], it was shown that the *efficiency function* (2.20) of the *optimal* $(n, k, S)$-*algorithm* can be expressed by using the following recurrent relation:

$$F_p(n,k) = F_p\left(n;\underbrace{0,0,\ldots,0}_{t},p_{t+1},\ldots,p_{k-1},p_k\right)$$

$$= \sum_{j=0}^{t} F_p\left(n-1;\underbrace{0,0,\ldots,0}_{j},p_{t+1}-1,\ldots,p_{k-1}-1,\right.$$

$$\left. p_k-1,\underbrace{p,p,\ldots,p}_{t-j}\right) \tag{2.21}$$

for the following initial condition:

$$F_p\left(1;\underbrace{0,0,\ldots,0}_{t},p_{t+1},\ldots,p_{k-1},p_k\right) = t+1. \tag{2.22}$$

In Refs. [6, 16] the authors analyze the *extreme special cases* of the *optimal* $(n;p_1,p_2,\ldots,p_t,p_{t+1},\ldots,p_{k-1},p_k)$-*algorithm* and their *efficiency functions* defined by (2.21) and (2.22) for these cases.

Let $p = 0$. Recall that for this case, all the *indicatory elements* can be enclosed to points of the segment $AB$ at each step of the algorithm and all the *indicatory elements* after the first step remain in the *state* of 0.

In Ref. [16], it is shown that for the case of $p = 0$, the *optimal* $(n; p_1, p_2, \ldots, p_t, p_{t+1}, \ldots, p_{k-1}, p_k)$-*algorithm* reduces to the *optimal* $(n, k, 0)$-*algorithm*, considered in Chapter 1, and the *efficiency function*, given by (2.21) and (2.22), reduces to the expression:

$$F_0(n, k) = (k + 1)^n, \tag{2.23}$$

in particular, to the expression

$$F_0(n, 1) = 2^n \tag{2.24}$$

for the case of $k = 1$ (the *classical binary algorithm*).

Now, let $p = \infty$ and the *initial conditions* of the *indicatory elements* (2.20) at the first step be equal:

$$p_1 = p_2 = \cdots = p_{k-1} = p_k = 0.$$

It is shown in [16] that for this case, the *optimal* $(n, k, S)$-*algorithm* reduces to the *optimal* $(n, k, 1)$-*algorithm* (the *binomial algorithm*), studied in Chapter 1, and its *efficiency function* (2.21) and (2.22) is given by the following expression:

$$F(n, k) = C_{n+k}^n = C_{n+k}^k. \tag{2.25}$$

Finally, let's consider the case of $k = 1$. It was shown in [16] that for this case, the *optimal* $(n, k, S)$-*algorithm* reduces to the *Fibonacci measurement algorithm*, discussed above, and its recurrent formula for the *efficiency function* (2.22) reduces to the expression (2.9).

The main result of the algorithmic measurement theory with all unexpected relationships is demonstrated in Table 2.3.

Thus, the "unexpectedness" of the obtained result consists in the fact that the solution of the generalized version of the *Bachet–Mendeleev problem* reduces to a very general recurrent relation for the *efficiency function* given by (2.21) and (2.22). This recurrent relation includes as a special case a number of well-known combinatorial formulas, in particular, the *combinatorial formulas for the number of arrangements with repetitions* $(k + 1)^n$, for the *number of combinations* $C_{n+k}^k = C_{n+k}^n$, for the *binary* $(2^n)$ and

Table 2.3. The main result of the algorithmic measurement theory.

| | $p = 0$ | $0 < p < \infty$ | $p = \infty$ |
|---|---|---|---|
| $k \geq 1$ | $(k+1)^n$ | $\leftarrow F_p(n, k) \rightarrow$ | $C_{n+k}^k = C_{n+k}^n$ |
| | $\downarrow$ | $\downarrow$ | $\downarrow$ |
| $k = 1$ | $2^n$ | $\leftarrow F_p(n) = F_p(n-1) + F_p(n-p-1) \rightarrow$ | $n+1$ |
| | Binary numbers | Fibonacci $p$-numbers | Natural numbers |

*natural* $(n + 1)$ *numbers*, as well as the *recurrent relation* (2.8) for the *Fibonacci p-numbers* and the recurrent relation for the *classical Fibonacci numbers*, which are given by (2.11) and (2.12).

## 2.7. Isomorphism Between "Lever Balances" and "Rabbits Reproduction"

### 2.7.1. What is isomorphism?

The concept of *isomorphism* [from the Greek words *isos* (equal) and *morphe* (form)] is widely used in mathematics. Informally, *isomorphism* is a kind of correspondence between objects, which show the relationship between their properties or operations. If there is *isomorphism* between two structures, we call these structures *isomorphic*. The word *isomorphism* is used when two complex structures can be reflected on one another in such a way that each part of one structure corresponds to a certain part of another structure.

We can give the following physical examples of *isomorphism*:

(1) **Wooden and metal cubes:** Although their physical nature is different, their geometric structures are *isomorphic*.

(2) **Big Ben in London and a wrist watch:** Although these clocks differ in size, their time-counting mechanisms are *isomorphic*.

In the 1976 article by Stakhov [57], the *principle of measurement asymmetry* was formulated. It should be noted that this principle has a practical origin and reflects some of the technical properties of the *comparators*, used in analog-to-digital converters for electrical quantities. This principle is based on the concept of the *inertial lever balances* used in weighing. If we denote through $I$ the *initial position* of the *lever balances* (Fig. 2.3(a)) and through $O$ the *opposite position* (Fig. 2.3(b)), then the operation of the *lever balances* can be described by using the two transitions:

$$I \to \begin{cases} I, \\ O, \end{cases} \tag{2.26}$$

$$O \to I. \tag{2.27}$$

The transition (2.26) means that the *lever balances* can pass to one of the two extreme positions $I$ or $O$, after we put the next standard weight on the free cup of *lever balances*. The transition (2.27) means the following. If the *lever balances* prove to be in the *opposite position* $O$, then within a certain time, the *lever balances* return back to the *initial position* $I$.

Now, in view of the foregoing, we will once again consider the classic *task of rabbits reproduction*, formulated by Fibonacci in the book *Liber abaci* (1202). Let's recall that the *law of rabbit's reproduction* reduces to the following. Each mature pair $A$ of rabbits produces a newborn couple $B$ within one month. The newborn couple becomes mature within one month and then starts producing one newborn couple each month. We can simulate the process of *rabbits reproduction* by using two transitions:

$$A \to AB, \tag{2.28}$$

$$B \to A. \tag{2.29}$$

Let's note, that the transition (2.28) models the process of *birth* of the new pair of rabbits $B$, and the transition (2.29) models the process of their *ripening*. At the same time, the transition (2.28) reflects the *asymmetry of rabbits reproduction* because according to (2.28), the mature pair $A$ is transformed into two non-identical pairs: the mature pair $A$ and the newborn pair $B$.

By comparing the transitions (2.26) and (2.27) with the transitions (2.28) and (2.29), we can see the analogy or the *isomorphism* between them. Moreover, this *isomorphism* is confirmed by the fact that solving these tasks, having a different physical nature (the *weighing process* and the *process of rabbits reproduction*), leads to the same recurrent numerical sequence: the *Fibonacci numbers!*

### 2.7.2. The generalized "principle of asymmetry of reproduction" in wildlife

By using the model of *rabbits reproduction*, which is described by the transitions (2.28) and (2.29), we can generalize the *task of rabbits reproduction* as follows. Let's give some non-negative number $p \geq 0$ and formulate the following task:

> *"Let there be the pair of rabbits (female and male) in the fenced place on the first day of January. This pair of rabbits produces the new pair of rabbits on the first day of February. Each newborn pair of rabbits becomes mature already through $p(p = 0, 1, 2, 3, \ldots)$ months and then in the next month gives birth to a new pair of rabbits. The question arises: how many pairs of rabbits will be in the fenced site through one year, that is, through 12 months from the beginning of rabbits reproduction?"*

It is clear that for the case $p = 1$, the generalized version of the *task of rabbits reproduction*, formulated above, coincides with the *classical Fibonacci task* formulated in the 13th century (see Chapter 2, Volume 1).

Note also that the case $p = 0$ corresponds to such an idealized situation, when the rabbits become mature immediately after *birth*. This case can be modeled by using the following transition:

$$A \to AA. \tag{2.30}$$

It is clear that the transition (2.30) reflects the *symmetry* of *rabbits reproduction*, when the mature pair $A$ turns into two identical mature pairs $AA$. It is easy to show that in this case, the *rabbits* are reproduced according to the *dichotomy principle*, that is, the number of *rabbits* double every month: $1, 2, 4, 8, 16, 32, \ldots$.

Now, we consider the case of $p > 0$. Let us analyze the new version of the *task of rabbits reproduction* with regard to the new rules of their *reproduction*, when the process of *reproduction* satisfies the more complex system of *transitions*.

Indeed, let $A$ and $B$ be the pairs of mature and newborn rabbits, respectively. Then the transition (2.29), as before, models the process of the monthly appearance of the newborn pair of rabbits $B$ from each mature pair $A$.

Let's now analyze the process of turning the newborn couple $B$ into the mature couple $A$. Because the process of *ripening* of the newborn couple $B$ now takes place during $p$ $(p = 0, 1, 2, 3, \ldots)$ months, this process includes the following intermediate stages corresponding to each month:

$$B \to B_1$$
$$B_1 \to B_2$$
$$B_2 \to B_3 \qquad (2.31)$$
$$\vdots$$
$$B_{p-1} \to A.$$

For example, for the case of $p = 2$, the process of turning a newborn couple $B$ into a mature couple $A$ consists of two intermediate stages:

$$B \to B_1, \qquad (2.32)$$
$$B_1 \to A. \qquad (2.33)$$

Then, taking into consideration the new "rules of the game", set by transitions (2.28), (2.32) and (2.33), the process of "rabbits reproduction" for the case $p = 2$ can be presented, as shown in Table 2.4.

Note that the column $A$ sets the number of mature pairs for each stage of reproduction, the column $B$ sets the number of newborn pairs, the column $B_1$ sets the number of newborn couples in stage

Table 2.4. "Rabbits reproduction" for the case $p = 2$.

| Date | Rabbit couple | $A$ | $B$ | $B_1$ | $A + B + B_1$ |
|------|---------------|-----|-----|-------|----------------|
| January 1 | $A$ | 1 | 0 | 0 | 1 |
| February 1 | $AB$ | 1 | 1 | 0 | 2 |
| March 1 | $ABB_1$ | 1 | 1 | 1 | 3 |
| April 1 | $ABB_1A$ | 2 | 1 | 1 | 4 |
| May 1 | $ABB_1AAB$ | 3 | 2 | 1 | 6 |
| June 1 | $ABB_1AABABB_1$ | 4 | 3 | 2 | 9 |
| July 1 | $ABB_1AABABB_1ABB_1A$ | 6 | 4 | 3 | 13 |

$B_1$, the column $A + B + B_1$ sets the total number of rabbits at each stage of reproduction.

Analysis of the numerical sequences in each column is given as follows:

$$A : 1, 1, 1, 2, 3, 4, 6, 9, 13, 19, \ldots,$$

$$B : 0, 1, 1, 1, 2, 3, 4, 6, 9, 13, 19, \ldots,$$

$$B_1 : 0, 0, 1, 1, 1, 2, 3, 4, 6, 9, 13, 19, \ldots,$$

$$A + B + B_1 : 1, 2, 3, 4, 6, 9, 13, 19, \ldots.$$

This shows that they obey one and the same regularity: each number $F_2(k)$ in the sequence is equal to the following sum: $F_2(k) = F_2(k - 1) + F_2(k - 3)$. However, as we know, the Fibonacci 2-numbers (see Chapter 4, Volume 1) obey this regularity.

By investigating this problem for the general case $p$ ($p = 0, 1, 2, 3, \ldots$), it is easy to conclude that the *Fibonacci p-numbers,* calculated according to the recurrent relation,

$$F_p(k) = F_p(k - 1) + F_p(k - p - 1)$$

are the solution of the generalized version of the *task of rabbits reproduction.* Thus, the above recurrent formula reflects the *generalized principle of asymmetry* that underlies the reproduction of living objects. Thus, the general formula reflects the *generalized principle of asymmetry,* which underlies the reproduction of living objects!

At first glance, it seems that the above *generalized task of rabbits reproduction* has no real sense. However, we shouldn't reach a hasty conclusion! Reference [67] is devoted to the application of the Fibonacci $p$-numbers for modeling the growth of biological cells. The article states as follows:

> "*In kinetic analysis of cell growth, the assumption is usually made that cell division, yields two daughter cells symmetrically. The essence of the semi-conservative replication of chromosomal DNA implies complete identity daughter cells. Nonetheless, in bacteria, yeast, nematodes and plants, the cell division regularly asymmetric, with spatial and functional differences between the two products of division.*"

In particular, this article [67] analyzes the models of cell growth based on the use of the Fibonacci 2- and 3-numbers. In this case, the authors of this article [67] make the following unexpected conclusion:

> "*Asymmetric division of the binary cells can be described by the generalized Fibonacci numbers... Our models, for the first time at the single cell level, provide rational bases for the occurrence of Fibonacci's and other recursive mathematical phyllotaxis and patterning in biology, founded on the occurrence of the regular asymmetry of the binary division.*"

Thus, the conclusions of this article [67] include the convincing confirmation of the fact that the processes of "reproduction" in Nature are subjected to the *generalized principle of asymmetry*, based on the *Fibonacci p-numbers*.

Now, let's return again to the *principle of measurement asymmetry*, formulated above, which led us to the *Fibonacci measurement algorithms*. We can see that the transition (2.32) in the *generalized task of rabbits reproduction* is similar to the return of *lever balances* to their initial position in the *Bachet–Mendeleev task*. This means that the *generalized task of rabbits reproduction* is *isomorphic* to the *Fibonacci measurement algorithm*!

Such *isomorphism* leads us to one more idea. Because the *principle of measurement asymmetry* underlies the main result of the *algorithmic measurement theory*, we can put forward the following hypothesis.

The algorithmic measurement theory [16], the main result of which is described by the very general recurrent relation (2.21) with the initial condition (2.22), is *isomorphic* to the *dynamic theory of biological populations*. This means that, thanks to the concept of *isomorphism*, we can use the *algorithmic measurement theory* [16] in the development of the *dynamic theory of biological populations*, which, at first glance, seems very far from the *algorithmic measurement theory* [16].

## 2.8. Isomorphism Between the Algorithmic Theory of Measurement and the Theory of Positional Numeral Systems

The main result of the *algorithmic measurement theory*, obtained above (Table 2.3), is of interest in itself, both for the *combinatory* and for the *number theory*, however, it can lead to deeper conclusions of a methodological character if we take into account the fact that *at its origin, the concept of number, which then became the basis of arithmetic, not only had a specific character, but was inseparable from the concept of measurement, which later became the basis of geometry* [133].

This statement emphasizes the connection between the conceptions of *numbers* and *measurement*. This gives us reason to say the following. The *algorithmic measurement theory* affects the foundations of mathematics, in particular, such a poorly developed branch of modern number theory as numeral systems. In Ref. [16], published in 1977, Alexey Stakhov wrote the following:

> *"Plato's scornful attitude towards "school" arithmetic and its problems, as well as the absence of any serious enough need in the creation of new numeral systems in the practice of computation (for several centuries fully satisfied by the decimal system, and in the recent decades by the binary system in digital computers), and can be as an explanation of the fact, why in number theory there was no due attention to the numeral systems and in this part it did not go far ahead in comparison to the period of their origin."*

From this point of view, the *algorithmic measurement theory* [16] is an unusual mathematical theory. Its unusualness lies in the fact that for the first time in the history of mathematics, it put the problem of *mathematical studying of the measurement algorithms*. In addition, each "optimal" measurement algorithm "generates" its own positional method of representing natural numbers. That is, there is *isomorphism* between the *optimal $(n, k, S)$-algorithms* and *positional numeral systems.*

For example, the *optimal $(n, k, 0)$-algorithms* are *isomorphic* to all the well-known positional numeral systems, including the *Babylonian system* with base 60, *decimal* and *binary* numeral systems. The *optimal $(n, k, 1)$-algorithms* are *isomorphic* to the *binomial* numeral systems, which are already moving into the field of practical applications. Finally, the *Fibonacci measurement algorithms* are *isomorphic* to the so-called *Fibonacci p-codes*, which could be used in modern computer and measurement technologies. These codes will be considered in later chapters.

From the above reasonings, the important role of the *algorithmic measurement theory* [16] in the development of mathematics and computer science is emphasized.

This means that the *algorithmic measurement theory* concerns one of the theoretically poorly developed systems in the theoretical sense of the theory of numbers, namely the **theory of numeral systems**. Thanks to the concept of *isomorphism*, *algorithmic measurement theory* becomes a powerful source for the development of the theory of numeral systems as an important direction of the *elementary number theory.*

On the other hand, new numeral systems, synthesized in the *algorithmic theory of measurement*, generate new computer arithmetic, which are of fundamental interest for modern computers.

## 2.9. International Recognition of the Algorithmic Measurement Theory

### 2.9.1. All-union conference "information-measuring systems 1973"

In the development of the *algorithmic measurement theory*, a number of events, which can be considered as certain "milestones" in its formation, can be distinguished. The concept of *Algorithmic Measurement Theory* was first introduced by Alexey Stakhov in 1973 in the speech "Algorithmic Measurement Theory", made by him at the All-Union Conference on Information-Measuring Systems (1973, Ukraine, Ivano-Frankivsk).

The high appreciation of this speech by the conference participants, in particular, by the outstanding Soviet expert in the "information technology" Prof. Fedor Temnikov, inspired Alexey Stakhov to further research in this direction and write the original scientific book on the *algorithmic measurement theory* [16].

### 2.9.2. Scientific trip to Austria

In 1976, fate gifted Stakhov a "happy incident": a 2-month scientific trip to Austria for the purpose of scientific internship at leading Austrian universities.

At that time, the international situation promoted the consolidation of scientific ties between the USSR and the countries of Western Europe. For the realization of the decision of the historic Helsinki meeting of the heads of European states (1975), an agreement was signed between USSR and Austria on scientific cooperation and exchange of scientific personnel, primarily doctors of science and professors.

Usually, professors from Moscow or Leningrad universities were chosen for such "prestigious" scientific trips. In this case, the peculiarity of the situation was the fact that there was a requirement for a doctor of computer science with a good knowledge of German language, which at that time was not very popular among the metropolitan computer professors. Therefore, a good knowledge of German language had played a decisive role in choosing Stakhov's candidature for the trip to Austria.

On January 8, 1976, a half-empty airliner had delivered Alexey Stakhov from Moscow to Vienna, where he was met at the airport by the representatives of the USSR Embassy in Austria. At enormous speed, the ambassadorial car brought Stakhov to the Embassy hotel. Then it was explained to him that the master of sports in auto racing (who was also a master of sports in karate) served as the driver of the Embassy and now the driver was preparing for a rally.

The next day, Stakhov was introduced to his Austrian "boss", Professor Richard Eier, Director of the Information Processing

Institute, Vienna Technical University. This institute became the main place of Stakhov's scientific work in Austria.

Then, the floor was given to Alexey Stakhov. In his autobiographical book, *Under the Sign of the "Golden Section": the Confession of the Son of Soldier of Student's Battalion* (2003), Stakhov colourfully described his visit to the University of Innsbruck:

> *"As is known, 1976 was the Olympic year and it was the Austrian city of Innsbruck that was elected as the site of the 12th Winter Olympic Games. Of course, I wanted to attend the Olympics, and Professor Eier provided me with this opportunity (although this was not part of the official program of my stay in Austria). I was settled in the apartment of one of the students of the University of Innsbruck.*
>
> *In Innsbruck, I was very well received by Professor Albrecht, Director of the Institute of Statistics of the University of Innsbruck. I visited his beautiful home, met his family. I was invited to the Christmas masquerade ball, which every year during the Christmas holidays a large family of Professor Albrecht arranged for themselves. The tradition of their large family was that every year at the beginning of the year the family gathered for a masquerade ball. At the same time, the family coordinated on the theme of the next masquerade ball, and then the whole year there was preparation for this family holiday.*
>
> *For this purpose, special masquerade costumes, corresponding to the theme of the holiday, were sewn. Coincidentally, in 1976 the holiday was called "Moscow Nights". Therefore, I played the role of "wedding general" at this masquerade ball. I gave a welcoming speech at the festival, presented to parents of Professor Albrecht slides with views of Moscow and a bottle of Stolichnaya Vodka. This caused a general delight. In memory of this wonderful holiday, I was photographed with the wife of Professor Albrecht, dressed in a fancy dress, against the background of a painting of St. Basil's Cathedral." (see Fig. 2.6.)*

From Innsbruck, Stakhov's path led him to the wonderful Austrian city of Graz, which was part of his scientific internship program. In Graz, Stakhov was received by the rector of the Graz Technical University. He introduced him to the activities of the computer center and the university's nuclear atomic reactor and the work of some departments of radio-electronic profile.

Fig. 2.6.  The masquerade ball in Innsbruck (February 1976).

Stakhov's most vivid impression from visiting the Graz Technical University was the acquaintance with the outstanding European mathematician, Professor Alexander Aigner, who at that time headed the Mathematical Institute of Graz Technical University. Professor Aigner was one of the largest specialists in the field of number theory. As it turned out, the Fibonacci numbers and the golden ratio were one of his mathematical hobbies.

This fact determined a particular interest of Alexander Aigner to Stakhov's scientific research. Immediately after the conversation with Stakhov, Aigner suggested to Stakhov to make a speech at the scientific seminar of the Mathematical Institute of the Graz Technical University with an exposition of the basic scientific results.

After Stakhov's speech, a small banquet in honor of this event took place in one of the Graz restaurants, which was attended by the university rector and leading university professors. Professor Aigner, in his toast in honor of the Soviet science, recalled the famous Soviet mathematician and academician Linnik, who provided scientific support to Aigner in the form of a review to his scientific work.

Immediately after this, Professor Stakhov turned to Professor Aigner with a request to give feedback on his scientific research. Prof. Aigner prepared such a review and sent it to Professor Eier in Vienna. Let's consider the excerpt from Aigner's review:

*"Original ideas of Prof. Alexey Stakhov from the University of Taganrog (USSR) in the field of algorithmic measurement theory and computer arithmetic are also of considerable interest from the point of view of theoretical arithmetic and number theory. The central idea of the work is to replace conventional binary arithmetic with arithmetic formed by Fibonacci numbers $1, 1, 2, 3, 5, 8, \ldots$. During Prof. Stakhov speech, as well as during a long personal conversation, I had the opportunity to get to know and appreciate the very valuable ideas of Prof. Stakhov."*

In the autobiographical book, mentioned above, Alexey Stakhov wrote the following:

*"This review by Austrian prominent mathematician made a profound impression on Professor Eier, and he immediately reported about this review to the leading Austrian experts in the field of cybernetics and computer technology, in particular, to the President of the Austrian Cybernetics society, Professor Trappel, and Austrian Computer Society President, Professor Zemanek. A few days later, Professor Eier happily informed me that I was being asked to give a scientific lecture at the joint meeting of the Computer and Cybernetics societies of Austria. I promised to think and give my answer in a few days. My time-out was due to the need to inform the Soviet Embassy about such a proposal. In fact, my lecture dealt with a fundamentally new direction in the field of computer technology, which at that time was not yet patented abroad.*

*I think that the Embassy would hardly have assumed the responsibility to give the permission for a Soviet scientist to deliver the scientific lecture without the appropriate scientific expertise, if not for one circumstance.*

*The fact is that on March 5, 1976, the "historic" 25th Congress of Communistic Party of Soviet Union was supposed to begin its work. In connection with this event, intensive anti-Soviet propaganda was launched in the West countries.*

*In particular, the famous West German magazine "Spiegel" had published an enormous article, entitled "If Lenin knew this!" The* article had quoted the examples of the corruption and embezzlement in the highest echelons of party power. From this article I first learned about the "activities" of Kunaev, Rashidov, Galina Brezhneva, her numerous lovers and diamonds, etc.

The article shocked me, and for the first time I thought about the fact, where our country is heading and what kind of society we will build, if charlatans and state criminals will be at the head of the country.

However, in connection with this propaganda campaign, all Soviet embassies received the directive to provide in organization of the performances of Soviet scientists, artists and athletes abroad.

After a long conversation with the adviser to the ambassador for science, I managed to convince him that my work has a purely mathematical, fundamental character, and that there are no special secrets in it. Taking the positive decision about my lecture, the adviser warned me that the further development of my direction in the USSR and my scientific career would largely depend on the reaction of the Austrian scientists to my lecture.

I did understood what I was risking, but the support of Professor Aigner had already oriented the Austrian scientists to a positive reaction. In addition, I spoke German quite well and therefore I looked at my report with optimism and I began intensive scientific and language preparation for the lecture.

My lecture was called "Algorithmic Measurement Theory and the Foundations of Computer Arithmetic". The announcement of the lecture reported the following:

"The methods of number representation can be considered as special methods or algorithms of measurement. Such an interpretation is the main idea of the present lecture.

Main scientific results:

- New scientific principle — "The asymmetry principle of measurement";
- Algorithmic theory of measurements;
- Generalization of Fibonacci numbers theory;
- "Fibonacci" binary arithmetic as the way to improve the information reliability of computer systems."

The lecture was scheduled for March 3rd. The information about the lecture was sent to all leading scientific centers of Austria and, as it turned out, of West Germany. That is why many leading scientists from Austria and other countries arrived at the lecture.

The representatives of the Austrian laboratory of the company IBM, which was headed by Professor Zemanek, President of the Austrian Computer Society, had attended on the lecture. In addition, the lecture was attended by representatives of the Soviet Embassy in Austria and Soviet probationers in Austria.

The lecture lasted about 1 hour and 10 minutes. Language preparing was quite good, because I never even looked at the written text of the lecture. By the reaction of the audience, I understood that the lecture was perceived "with a bang". The next day, Professor Eier handed me the letter of the President of the Austrian Cybernetic society, Professor Trappel, in which he asked me to present the lecture manuscript for publication in the Proceedings of the Austrian Cybernetic society.

A wonderful review on the lecture was written by Professor Adolph Adam from the Linz University of Johannes Kepler. From this review I learned for the first time that the ingenious Kepler was fascinated with Fibonacci numbers and the golden section, and that his "speculations" on this subject are stored in the Pulkovo Observatory (near Leningrad).

But the most informative review I had received from my Austrian "chief" Professor Eier. By handing his review, he informed me that the night he had not sleep under the impression from my lecture. He concluded his review with the following words:

"Your clear and methodically impeccable report caused me great interest in your lecture, and I will try to carry out the relevant work in this field at my institute in the future. I hope that after returning to Taganrog our contacts will continue."

The next day I was invited by the ambassador's adviser in science. He conveyed to me the gratitude of Ambassador Efremov for the worthy presentation of Soviet science in Austria. He also informed me that the Embassy intends to inform the relevant authorities in the USSR about the reaction of Austrian scientists to my lecture.

On March 5, 1976, I flew from Vienna to Moscow, full of energy, hopes and new scientific ideas.

## 2.10. Letter of the USSR Ambassador to Austria, Mikhail Efremov, on Patenting Fibonacci Inventions

After returning to Taganrog, Stakhov plunged into the educational process, scientific work and family affairs, and began to gradually forget about Austria. A month later, he began to think that the promise of the Embassy adviser about scientific support was a polite diplomatic "empty chatter".

Soon, the Austrian events again resurfaced themselves. Somewhere in May 1976, either the Rector or Vice-rector of the Taganrog Radio Engineering Institute received a phone call from Moscow with a formidable message that the auditor was coming from Moscow to Taganrog in connection with the letter of the USSR Ambassador in Austria about the stay of Professor Stakhov in Austria. From such a phone call, the institute's administration came to the conclusion that Stakhov had done something very serious in Austria if the Ambassador himself had to write a letter to the appropriate authorities.

In any case, the Taganrog department of the State Security Committee was informed about the call from Moscow, and before the arrival of the Moscow guest, a total verification of Stakhov's activities at the Taganrog Radio Engineering Institute began (first of all, of his scientific and invention activities).

The Moscow guest turned out to be an expert of the patenting department of the USSR State Committee for Inventions and Discoveries, Agapov Viktor Mikhailovich, who handed Stakhov the copy of the letter from the USSR Ambassador in Austria, Mikhail Efremov, dated April 2, 1976. In his letter, the Ambassador of the USSR in Austria gave a detailed analysis of Stakhov's stay in Austria and especially of Stakhov's lecture at the joint meeting of the Cybernetics and Computer Societies of Austria.

The most important part of the Ambassador's letter turned out to be the following quote:

> *"Prof. Stakhov's research concerns to the foundations of digital computer arithmetic. In this direction, he managed to build an original theory of new binary numeral systems (the so-called "Fibonacci" numeral systems), which, compared to the classical "binary system", have a new quality: a high ability to detect errors in computers.*
>
> *This theory opens up new perspectives in enhancing of the informational reliability of digital computers and creates prerequisites for designing highly reliable self-checking digital computers. This part of the report aroused particular interest*

among Austrian scientists, involved in designing digital computers. At present, in Austria, the task of increasing the informational reliability of digital computers (especially special-purpose digital computers) is actual task and there is every reasons to assume that in Austria (in particular, in the IBM laboratory) there will start to implement the Fibonacci numeral systems.

In order to consolidate the priority of Soviet science in this direction, we consider it expedient to do the following proposals:

1. In view of the expressed interest among Austrian scientists in Prof. Stakhov's invention in the field of Fibonacci's numeral system (designing self-checking digital computers), it would be advisable to speed up the process of patenting Prof. Stakhov's invention what will preserve the priority of Soviet science and, possibly, will bring an economic effect.
2. Prof. Stakhov A.P. (through Prof. Albrecht, University of Innsbruck) managed to establish scientific contacts with a number of the leading German scientists in the field of computer science. Apparently, it would be advisable to promote to the further Prof. Stakhov's contacts with specialists from Austria and Germany."

Of course, this letter removed all suspicions of Stakhov, but the Rector of Taganrog Radio Engineering Institute was busy with the turbulent activity of "*revealing the anti-Soviet activities of Professor Stakhov abroad*". The Rector did not find time to meet with the expert from the Soviet Committee on Inventions and Discoveries. This fact, however, left an unpleasant aftertaste in Stakhov's soul and became one of the reasons for his subsequent decision to move to Vinnitsa (Ukraine).

What are the results of patenting Stakhov's inventions abroad? Stakhov's name is the first in the description of 65 foreign patents issued by the patent offices of the United States, Japan, England, France, FRG, Canada and other countries. This testifies, first of all, that the idea of Fibonacci computer is completely new and original, and Stakhov's patents are official legal documents confirming Stakhov's priority (and the priority of the now non-existent USSR state) in the new direction in this field of computer technology.

A high appreciation of Stakhov's inventions was given by the patent attorney of the USSR in Japan, who, during his visit to Moscow (1980) in his speech at the Chamber of Commerce and Industry of the USSR, noted global originality and perspective of Stakhov's inventions.

*Introduction to the Algorithmic Measurement Theory* **(1977):** A landmark event in the widespread acceptance of "algorithmic measurement theory" was the publication of Alexey Stakhov's book, *Introduction to Algorithmic Measurement Theory.* The book was published in 1977 by the Soviet Radio publishing house [16]. Around 25 years after the publication of this book, Stakhov not only did not abandon the basic ideas, set forth in the book, but also once again became convinced in the correctness of these ideas that led him in recent years to the concept of new mathematics, *The Mathematics of Harmony* [6], based on the Fibonacci numbers, golden section and algorithmic measurement theory.

*The 8th Congress of the International Confederation for Measuring Equipment and Instrumental Engineering:* In May 1979, the 8th Congress of the *International Confederation for Measuring Equipment and Instrumental Making* took place in Moscow. This congress was a major event in the field of measuring equipment and metrology, and outstanding scientists from around the world took part in its proceedings.

The plenary session of the Congress was held on May 21, 1979 in the famous Cinema Concert Hall "Russia". The lecture "Theoretical and physical-metrological problems of the further development of measuring equipment in industry and science" was the main event of the plenary session. Prof. Yu.V. Tarbeev (Russia) and Prof. D. Hoffman (F. Schiller University, Jena, GDR) were the authors of this lecture.

The lecture was delivered by Prof. Hoffman. A complete surprise for Stakhov was the repeated mention in the lecture of the *algorithmic measurement theory* as one of the new directions in theoretical metrology. By analyzing the modern theory of measurement, the authors of the above lecture emphasized that *in recent years, various versions of the general theory of measurements have been proposed: informational theory of measurements, quantum-mechanical (physical) theory of measurements, set-theoretical theory of measurements, algorithmic theory of measurements, and others.*

For Stakhov, this lecture was very special because the lecture was essentially the first official recognition of the *algorithmic theory of measurements* at such a high international level.

**Lecture at the International Symposium "Intellectual Measurements":** In 1986, Alexey Stakhov delivered the lecture *Algorithmic Theory of Measurements* at the plenary session of the *International Symposium on Intellectual Measurements*. This Symposium was held in Jena, DDR. The famous Soviet scientist Prof. Vladimir Kneller, who at that time was the Chief Editor of the journal *Measurement, Control, Automation*, attended Stakhov's lecture. He immediately suggested to Stakhov to write the article on the *Algorithmic Theory of Measurements* for his journal, and the article was published in this prestigious journal soon after the Symposium [60].

The article in the International Journal *Computers & Mathematics with Applications*: In 1989, the international journal *Computers & Mathematics with Applications* published Stakhov's great article "The Golden Section in the Measurement Theory" [61]. This was Stakhov's first publication in English. It should be noted that this journal has always enjoyed and still enjoys high prestige in the scientific community. The reviewers of the articles are world famous scientists, and therefore, the publication of Stakhov's article [61] in this journal also pointed to the international recognition of the algorithmic measurement theory.

The article in the academic journal *Control Systems and Machines*: In 1994, the international journal *Control Systems and Computers* (Academy of Sciences of Ukraine) published Stakhov's article "Algorithmic Measurement Theory: A New View at the Theory of Positional Numeral Systems and Computer Arithmetic's" [65].

To emphasize the significance of this publication, the editorial board put the title *Algorithmic Measurement Theory* on the front cover of the journal as one of the main topics of the issue along with such topics as "Informatization of Ukraine" and "New Methods in Informatics".

Thus, without any doubts, the concept of algorithmic theory of measurements, introduced by Stakhov in 1973, was widely included in the modern scientific and technical literature, and the *algorithmic theory of measurements* [16] is now recognized as the original scientific direction of modern mathematics, informatics and theoretical metrology.

# Chapter 3

# Evolution of Numeral Systems

## 3.1. Brief History of the Most Known Numeral Systems

### 3.1.1. The main stages in the development of numeral systems

The creation of the first numeral systems relates to the period of the origin of mathematics, when the need for counting objects and the amount of foodstuff, measuring of time and land led to the creation of the simplest concepts of arithmetic of natural numbers. In Ref. [102], the prominent Russian mathematician A.N. Kolmogorov notes that *"only on the basis of the oral numeral systems the written numeral systems arise and the methods of performing four arithmetic operations over natural numbers are gradually developed."*

There are several stages in the history of numeral systems, such as *initial stage of counting, non-positional numeral systems, alphabetic systems, positional numeral systems.*

The initial stage of counting is characterized by the image of the counted objects by using body parts, especially fingers of hands and feet, sticks, knots of rope, etc. As emphasized in Ref. [133], *"despite the extreme primitiveness of this mode of representation of numbers, it played an exceptional role in the development of the concept of a number."* Thus, it was the counting system that played a decisive role in the development of the *concept of a natural number*, one of the fundamental concepts of mathematical science.

Unfortunately, historians of computer science, who have dealt with the developed computer theory, sometimes forget about the role played by the numeral systems in the history of computer science. Indeed, the first primitive projects of computing devices (abacus and arithmometers), the prototypes of modern computers, appeared long before the emergence of algebra of logics, information theory and algorithms theory. It is important to note that the numeral systems played a decisive role in the creation of the first primitive prototypes of modern computers. This should not be forgotten while predicting the further development of computer science.

### 3.1.2. Babylonian numeral system with the base 60

Already in the period of the original mathematics, one of the greatest discoveries in the field of number systems was made: the *Babylonian positional principle of number representation.* As emphasized in Ref. [133], *"the first known numeral system, based on the positional principle, was the ancient Babylonian numeral system with the base of* 60, *which emerged about* 2000 *years BC."*

To explain the question of the origin of base 60, several competing hypotheses arose in the history of mathematics [133]. According to some researchers, base 60 of the Babylonian numeral system is connected with the numbers 12 and 5 by the following relation: $60 = 12 \times 5$, where 5 is the number of fingers on the hand. There is also the hypothesis of Neugebauer [134] that after the Akkadian conquest of the Sumerian state, the two monetary and weight units existed there for a long time: *shekel* and *mine*, and their relation was as follows: 1 *mine* = 60 *shekels*. Later, this relation became customary and gave rise to the corresponding system of recording of all numbers.

The well-known historian of mathematics M. Cantor initially suggested that the Sumerians (the primary population of the Euphrates valley) had considered the year to be equal to 360 days and that the system with base 60 has an *astronomical origin.*

According to the hypothesis of another historian mathematician G. Kiewich, the two nations met in the Euphrates valley; the first nation used the *decimal system* with base 10, and the other one used the *numeral system with base* 6 (the appearance of base 6 is explained by Kiewich with the special calculation on fingers, where the hand, clenched into the *fist*, meant 6). Thanks to the merger of both the systems, the "compromise" base 60 = 6 × 10 had emerged.

The Babylonian state also inherited the positional system with base 60 and passed it, along with the sky observation tables, to the Greek astronomers. At a later time, the Babylonian system with base 60 was used by Arabs, as well as by the ancient and medieval astronomers, primarily to represent fractions, which were often called the *"astronomical" fractions*.

In the 13th century, the influential rector of the University of Paris, Peter Filomen, advocated the widespread introduction of the system with base 60 in Europe. In the 15th century, Johann Gmunden, Professor of Mathematics at the University of Vienna, made a similar appeal. Both initiatives did not yield the expected results.

From the beginning of the 16th century, the *decimal representation of the fractions in Europe* was completely replaced by their *astronomical representation in Babylonian system with base 60*. Now, the remnants of the Babylonian system are used only in the measurement of *angles* and *time*.

Note that the above hypotheses of Neugebauer, Cantor and Kiewich concern, first of all, the question about the origin of base 60, but not the positional principle of number representation. The last question is answered by another *Neugebauer hypothesis*, which concerns the origin of the *positional principle*. According to this hypothesis [134], *"the main stages of the formation of the positional system in Babylon were:* (1) *establishing a quantitative relationship between two independent existing systems of measures and* (2) *lowering the names of the discharge units when writing."*

Neugebauer considered that these stages of the origin of positional numeral systems were general for all the ancient cultures; he emphasized that *"the positional system with the base* 60 *turned out to be a completely natural final result of the long historic development that is not fundamentally different from similar processes in other cultures"*.

By the way, *Neugebauer's hypothesis* is in good agreement with the idea of *isomorphism* between the *algorithmic measurement theory* [16] and the *positional numeral systems*. According to this idea, the *algorithmic measurement theory* [16] not only generates all known positional numeral systems (*with base* 60, *decimal, binary, ternary* and so on), but is also the source of the origin of the *fundamentally new numeral systems* (in particular, *Fibonacci and binomial numeral systems*).

In Volume I (Chapter 4) of the present three-volume book, we discuss a new view on the ancient *Egyptian calendar* and all systems of time and angular quantities and connection of their numerical characteristics 12, 30, 360 as their *nodal numbers* (1 year = 12 months, 1 month = 30 days, 1 day = 24 = 2 × 12 hours, 1 hour = 60 minutes, 1 minute = 60 seconds) with numerical parameters of the *dodecahedron*. In Chapter 4, we analyzed the relationship of the numerical characteristics of *the dodecahedron* 12, 30, 60 and 360 = 12 × 30 with the numerical characteristics of the Egyptian calendar and modern systems for measuring *time and angular values*. It is possible that the *dodecahedron*, which in the ancient science symbolized the *Harmony of the Universe*, was the reason for choosing the number 60 (the number of flat angles on the surface of dodecahedron) as the basis of the Babylonian numeral system with base 360.

### 3.1.3. Alphabetical and Roman numbering systems

The so-called *alphabetical systems of numeration* were widely used in many nations, when the letters of the alphabet were assigned certain numerical values. The *alphabet systems of numeration* were used by the ancient Armenians, Georgians, Greeks (Ionic numeral system), Arabs, Jews and other people of the Middle East.

In the Slavic liturgical books, the Greek alphabet system was translated into Cyrillic letters. For example, the so-called *beastly number* of 666 is associated with the *alphabetic numeration*. If we write two words *Emperor Nero* in Hebrew and then calculate the sum of the numerical values of the letters, which this phrase contains, then this sum will be equal to 666. Thus, the number 666 was a hidden designation of the name of Nero, the persecutor of Christianity, the *man-beast*, who with extreme cruelty dealt with his political opponents and with whose consent his mother, both wives, the philosopher Seneca and many others famous persons were killed.

Slavs also used alphabetic numeration. But in order to distinguish letters from numbers, they were writing a special sign called *titlo* above the letters, depicting the numbers.

Everyone knows the *Roman numbering system*, which preceded the appearance of our positional numeral system. But the *Roman system was not positional* (more precisely, it was partially positional system). It uses Latin letters as numbers: $I$ meant 1 $(I = 1)$, $V = 5$, $X = 10$, $L = 50$, $C = 100$, $D = 500$, $M = 1000$. For example, $II = 1 + 1 = 2$; here, the symbol $I$ denotes 1 regardless of the place in the number. The elements of the "positionalness" of the *Roman numeral system* consisted in the fact that for some records of numbers, the meaning of the numbers depended on their position in this record. For example, $IV = 5 - 1 = 4$, while $VI = 5 + 1 = 6$. A significant drawback of the *Roman numeral system* was the fact that it was not adapted to perform arithmetic operations in the writing form.

### 3.1.4. Ancient Egyptian decimal non-positional numeral system

The *ancient Egyptian decimal numeral system* appeared in the second half of the third millennium BC. It was the *hieroglyphic* system. The Egyptian numbering system was *decimal* but *non-positional*. Numbers had been presented as the sum of some *nodal numbers*: $1, 10, 10^2, 10^3, 10^4, 10^5, 10^6, 10^7$. Special hieroglyphic signs were used for designating *nodal numbers*, which played the role of the *numerals*. For example, in the record of the number $325 = 3 \times 100 + 2 \times 10 + 5 \times 1$, Egyptians used the following hieroglyphs: the five hieroglyphs, which represented the *nodal number* of 1, the two hieroglyphs, which represented the number of 10, and the three hieroglyphs, which represented the number of 100.

The main achievement of the Egyptian arithmetic was the invention of the *doubling principle,* which allowed the Egyptian mathematicians to perform the operations of *multiplication* and *division* of numbers. In order to multiply number 35 by 12, the Egyptian mathematician acted as follows. He had constructed a table of three columns. The binary numbers $2^k (k = 0, 1, 2, 3, \ldots)$ was placed in the first column of the table; in the second column, the first number was one of the multipliers (in this case, 35), and each subsequent number was equal to the double of the previous number:

| 1 | 35 | |
|---|---|---|
| 2 | 70 | |
| /4 | $\to 140 \to$ | 140 |
| /8 | $\to 280 \to$ | 280 |
| | $35 \times 12 =$ | 420 |

Then the binary numbers of the first column, the sum of which was equal to the second multiplier ($12 = 4 + 8$), were marked with a *slash*. The result of multiplication was obtained by the summation of the numbers of the second column, corresponding to the marked numbers of the first column. The analysis of the *Egyptian multiplication method,* based on the *doubling principle* leads us to the much unexpected conclusion. The decomposition of the binary integer of $2^k (k = 0, 1, 2, 3, \ldots)$, used in the first column, is nothing more than its representation in the binary numeral system ($12 = 1100$). On the other hand, if the multiplier 35 is represented in the binary system ($35 = 100011$), then the doubling of number 35, used in the second column and performed at each multiplication step, can be done by shifting the binary code of the number 35 by one digit to the left ($70 = 1000110$, $140 = 10001100$, etc.). In other words, the Egyptian multiplication method, based on the *doubling principle,* essentially *coincided with the basic algorithm of number multiplication in modern computers!*

The "doubling principle" was used for integer division. If, for example, we need to divide the number 30 by 6, then the Egyptian mathematician did it as shown in the following table:

| /1 | $6 \leq 30$ | $6 \leq 6$ |
|---|---|---|
| 2 | $12 \leq 30$ | $12 > 6$ |
| /4 | $24 \leq 30$ | |
| 8 | $48 > 30$ | |

The binary numbers $2^k (k = 0, 1, 2, 3, \ldots)$ are written in the first column; in the first row of the second column, the divisor 6 was recorded, and each subsequent number of the second column was doubled (6, 12, 24, 48, and so on). At each step of "doubling", the numbers of the second column were compared with the divisible 30 until the next "doubled" number (in this case, the number 48) had turned out strictly *more* than the divisible number 30 ($48 > 30$). After that, the doubled number of the previous column had been subtracted from the divisible number 30 ($30 - 24 = 6$), while the binary number 4 of the first column had been marked with a *slash*. Then with the difference ($30 - 24 = 6$), the same procedure had been done as in the second column, that is, the divisor 6 and all the "doubled" numbers were compared with the difference of 6 until the "doubled" number became strictly *more* than the difference of 6 ($12 > 6$). After marking the number 1 of the first column by a *slash* and subtracting the divisor of 6 from the difference of 6, we get the number $0 (6 - 6 = 0)$ or the integer *less* than the divisor. At this, the division ended. The sum of the "binary" numbers of the first column, marked with a *slash* $(4 + 1 = 5)$, was the result of the division.

> *"Thus, we should admire the genius of the Egyptian mathematicians, who many thousands of years ago invented the methods of multiplication and division of numbers that are used in modern computers!"*

### 3.1.5. Mayan numeral system

The origin of *positional numeral system is considered as one of the main achievements in the history of material culture*. Many nations took part in its creation. In the sixth century, the positional numeral system originated in the *Mayan tribe*. The most common opinion is the fact that the base of the *Mayan numeral system* was the number $20 = 10+10$, which meant that the *Mayan numeral system* had *finger* origin. However, it is known that in the *Mayan numeral system*, there was only one deviation from base 20. The weight of the next *nodal number*, following the *nodal number* of 20 was chosen by the Mayans to be 360 (not $20^2 = 400$). All subsequent weights of the digits are derived from 20 and 360 and they are the *nodal numbers* that make up the *Mayan numeral system*. As emphasized in Ref. [133], this *is explained by the fact that the Mayans divided the year into 18 months, 20 days each, plus the five more days.*

This means that the *nodal numbers* of the *Mayan numeral system* were of *astronomical origin*. Given the high level of development of the *Mayan culture*, it can be assumed that the Mayans were familiar with the *Platonic solids* and that their numeral system was associated with the *icosahedron*, dual to the *dodecahedron*.

### 3.1.6. Decimal system

We use the decimal number system for everyday calculations. The oldest known record of the *positional decimal system* was discovered in India in the sixth century. The greatest invention of the Indians was the introduction of *zero*. Recently, the American mathematician, Charles Safe, published the book *Biography of the Number Zero*. He wrote the following about *zero*:

> "In the number of zero, there is a hint of the indescribable and the ineffable; it contains the unlimited and the infinite. That is why, it had been feared, hated, and even forbidden."

For millennia, people had lived without *zero*: this number was unknown to neither the Egyptians, nor the Romans, nor the Greeks, nor the ancient Jews. Only in India, zero finally took its worthy place in mathematics, and then the Indian numeration with the number of *zero* spread throughout the world. The Indian numbering came first to the Arab countries and later to Western Europe.

The Central Asian mathematician Al-Khwārizmī spoke about the Indian numbering to the Arab world. Simple and convenient rules for adding and subtracting numbers, written in the Indian positional system, made it especially popular. As Al-Khwārizmi wrote his mathematical work in Arabic, the Indian numbering in Europe became known under the title *the Arab system*, which is wrong from a historical point of view.

In Europe, the *decimal numbering penetrated from the Islamic East.* The earliest Arabic manuscripts, containing the *Indian positional numeral system*, date back to the 9th century AD. The French clergyman and mathematician Herbert, who in 999 became the Roman Pope under the name of Sylvester II, was one of the first in Europe who understood the advantages of the new numbering. Pope Sylvester II tried to reform the teaching of mathematics based on a new numbering system. However, the innovation was met with fierce anger from the side of the Inquisition. The Pope was accused of *selling his soul to the Saracen devils.* The reform failed, and the Pope mathematician soon died. But after his death, for several centuries, it was rumoured that smoke was constantly trickling out from the Pope's marble sarcophagus and sounds of devils were heard. Although the first Arabic–Indian numerals were found in the *Spanish manuscripts* as early as the 10th century, the *decimal system began to spread in Europe, by the beginning of the* 12*th century.* The new numbering in Europe met with resistance both from the official scholastic science of that time and from individual governments. For example, in 1299, in Florence, it was *forbidden* for merchants to use the *decimal numeration,* while in bookkeeping, it was ordered to either use Roman numerals or write numbers in words. The famous Italian mathematician Leonardo from Pisa (Fibonacci), who received a mathematical education in the Arab countries, was a strong supporter of the use of the Arabic–Indian numeral system in the trading practice. In his book *Liber abaci* (1202), he wrote as follows:

> *"The nine Indian signs are the following:* 9, 8, 7, 6, 5, 4, 3, 2, 1. *With the help of these signs and the* 0 *sign, which is called "zephirum" in Arabic, we can write any number you like."*

Here, the word *zephirum* meant the same as the Arabic word *as-sifr*, which is the literal translation of the Hindu word of *sunya*, that is, the *empty word*, which served as the name of *zero*. The word *zephirum* gave rise to the French and Italian word *zero*. On the other hand, the same Arabic word *as-sifr* was transmitted through *ziffer*, from which the French word *chiffre*, the German word *ziffer*, the English word *cipher* and the Russian word **цифра** originated.

At the beginning of the 17th century, the decimal numeration penetrated into Russia, but the Orthodox Church received it with enmity and announced the decimal system is witchcraft and godless. The decimal system came into usage in Russia only after the publication of the famous book *Arithmetic* by Magnitsky (1703), in which all calculations in the text were made exclusively by the usage of the decimal numeral system.

There are many opinions about the choice of number 10 as the base of the decimal numeral system. The most common is the opinion that the base of the decimal system had the *finger* origin. However, one should not forget that number 10 always carried a special semantic load in ancient science and had a relation to *Universal Harmony*. The Pythagoreans called this number as the *tetraxis*. The *tetraxis* $10 = 1 + 2 + 3 + 4$ had been considered by the Pythagoreans as one of the highest values and was *a symbol of the whole Universe* because it contained the four *basic elements*: The *unit* or the *monad* of 1, denoted, according to Pythagoras, as the *spirit*, from which the entire visible world flows. The *deus* or *dyad* $(2 = 1 + 1)$ symbolized the *material atom*; the *triple* or *triad* $(3 = 2+1)$ was the symbol of the *living world*. The *four* or *tetrad* $(4 = 3+1)$ connected the *living world* with the *monad* and therefore symbolized the *visible* and *invisible* worlds, and because the tetraxis $10 = 1 + 2 + 3 + 4$, it expressed the *Whole*. Thus, the hypothesis of the *harmonious origin* of the tetraxis of 10 has *no less* the right to exist, just like the hypothesis of the *finger origin*.

### 3.1.7. The decimal system as the greatest mathematical discovery in the history of mathematics

Every person on the globe, who has graduated at least four classes of the *elementary* or *parish* schools, must know at least two useful things: he must write and read using the decimal system to perform the simplest of arithmetic operations. This system seems to be so *simple* and *elementary* that many of us will treat it with much mistrust to the statement that the *decimal system is one of the greatest mathematical discoveries in the history of mathematics*. In order to convince the reader on this, let us turn to the opinions of the authoritative experts in the field.

**Pierre Simon Laplace** (1749–1827), French mathematician, member of the Paris Academy of Sciences, the honorary foreign member of the St. Petersburg Academy of Sciences:

> *"The thought to express all the numbers with 9 characters, by giving them, apart from the value by form, another value by place, is so simple that it is difficult to understand just how it is surprising. How it was not easy to come to this method, we see on the example of the greatest geniuses of Greek scholarship of Archimedes and Apollonius, from whom this thought remained hidden."*

**M. V. Ostrogradsky** (1801–1862), Russian mathematician, member of the St. Petersburg Academy of Sciences and many foreign academies:

> *"It seems to us that after the invention of written language, the biggest discovery was the use of the so-called decimal numeral system. We want to say that the agreement, according to which we can express all the useful numbers by twelve words and their endings is one of the most remarkable creations of human genius."*

**Jules Tannery** (1848–1910), French mathematician, the member of the Paris Academy of Sciences:

*"As for the current written numbering system, which uses nine significant ciphers and zero, and the relative significance of numbers is determined by a special rule, this system was introduced in India in the era that is not defined precisely, but apparently after the Christian era. The invention of this system is one of the most important events in the history of science, and despite on the habit of using decimal numbering, we cannot be not surprised at the marvellous simplicity of its mechanism."*

**The 12-decimal notation** is a positional numeral system with base 12. The numbers 0, 1, 2, 3, 4, 5, 6, 7, 8, 9, A, B are used as numerals of this numeral system. Number 12 could be a very convenient base of this numeral system because it is divided completely into 2, 3, 4, and 6, while number 10, the base of the decimal numeral system, is divided completely into only 2 and 5.

The 12-decimal notation originated in ancient Sumer. Some people of Nigeria and Tibet are using the 12-decimal notation at the present time. The 12th fractions were often found in the European systems of measures. The Romans had a *standard fraction* of the *ounce* (1/12). 1 *English penny* = 1/12 *shillings*, 1 *inch* = 1/12 *feet*, etc.

The usage of the 12-*decimal* notation in the calculation practice had been proposed repeatedly. In the 18th century, the famous French naturalist Buffon was its supporter. During the Great French Revolution, a "Revolutionary Commission on Weights and Measures" was established, which for a long period considered a similar project, but due to the efforts of Lagrange and other opponents of the reform, the project was curtailed. The main argument against this included the huge costs and the inevitable confusion during the usage.

### 3.1.8. Binary system

A history of the *binary system* is rather curious. Many nations and outstanding scientists from different countries and continents took part in its creation. Let's start with the famous Chinese book, *Book of Changes*, the most significant work of the ancient Chinese philosophy. The existing system in the *Book of Changes* was formed mainly during the *Zhou dynasty* (1046–256 BCE). It consisted of 64 characters: *hexagrams*, each of which expresses a particular life situation in time in terms of its gradual development. A complete set of eight *trigrams* and 64 *hexagrams*, which is essentially the equivalent of the 3-bit and 6-bit digits, were known in ancient China in the classic texts of the *Book of Changes*. The order of hexagrams in the *Book of Changes*, arranged in accordance with the values of the corresponding binary digits (from 0 to 63), and the method of obtaining them was developed by the Chinese scientist and philosopher Shao Yong in the 11th century. However, there is no evidence that Shao Yong has developed the rules of the binary arithmetic.

One of the first mentions of the *binary system* is found in the works of the Indian poet and mathematician Pingala (200 BC), who developed the mathematical foundations for describing poetry by using the binary system.

In 1605, the famous English scientist and lawyer Francis Bacon (1561–1626) described a system, whose letters of the alphabet can be reduced to a sequence of the binary numbers (Fig. 3.1). Bacon was a prominent scientific figure during the *Renaissance* and *Scientific Enlightenment*. In particular, Bacon developed and popularized the scientific method that marked a new scientific rigor, based on the evidence, results and the methodical approach to science.

Francis Bacon is widely considered to be the father of empiricism and the Scientific Revolution of the Renaissance period. He made

Fig. 3.1. Francis Bacon (1561–1626).

Fig. 3.2. Wilhelm Gottfried Leibniz (1646–1716).

an important step in the development of a general theory of binary coding, drawing attention to the fact that this coding method can be applied to the arbitrary object.

The great European, German scientist Wilhelm Gottfried Leibniz (1646–1716) is considered as the creator of the *binary arithmetic* (Fig. 3.2). From his student years to the end of his life, he studied the properties of the *binary system*, which later became the basis for the creation of modern computers. The binary system was fully described by Leibniz in the 17th century in the work, *Explication de l'Arithmétique Binaire*. He had attributed to the binary arithmetic a certain mystical meaning and believed that at its basis it was possible to create a universal language for explaining the phenomena of the world and for using it in all sciences, including philosophy.

Leibniz made outstanding contributions to designing of the first mechanical calculators. As noted in Wikipedia, "*he became one of the most prolific inventors in the field of mechanical calculators. While working on adding automatic multiplication and division to Pascal's calculator, he was the first who describes a pinwheel calculator in 1685 and invented the Leibniz wheel, used in the arithmometer, the first mass-produced mechanical calculator. He also refined the binary numeral system, which is the foundation of all the digital computers.*"

The image of the medal, made by Leibniz in 1697 is preserved and it explains the relationship between the *binary* and *decimal* numeral

Fig. 3.3.  George Boole (1815–1864).

systems. The medal represented a plate of two columns: in the first column, the integers 0–17 in the decimal system are represented, and in the second, the same integers in the binary system are represented. As a person, fascinated with Chinese culture, Leibniz knew about the *Book of Changes* and was one of the first scientists who noticed that hexagrams correspond to the binary numbers from 0 to 111111. He believed that the *Book of Changes* was an evidence of the major Chinese achievements in the philosophical mathematics of the time.

Leibniz did not recommend the *binary system* in place of the *decimal* system for the practical calculations, but he had stressed [133] that *"computations by using 0 and 1, in reward for his lengthways is the main for science and even in the practice of numbers, and especially in geometry: the reason for this is the fact that when reducing numbers to the simplest principles, what are 0 and 1, a wonderful order is revealed"*. With this statement, he anticipated the modern "computer revolution" based on the *binary system*.

**Boolean algebra:** In 1854, the English mathematician George Boole (1815–1864) published the landmark work, *The Laws of Thought*, which described the algebraic systems applied to logic, which is currently known as *Boolean algebra* or *algebra of logic*. Its logical calculus was destined to play an important role in the development of the modern digital electronic circuits (Fig. 3.3).

Fig. 3.4.  Claude Shannon (1916–2001).

**Claude Shannon Studies:** Claude Elwood Shannon (1916–2001) is the American mathematician, electrical engineer, and cryptographer, known as *the father of information theory*, described by him in his 1948 paper *Mathematical Theory of Communication*. He became, perhaps, well known widely for the creation of the *digital circuit design theory* (1937), when — as a 21-year-old master's degree student at the Massachusetts Institute of Technology (MIT) — he wrote his PhD thesis devoted to the electrical applications of the Boolean algebra. Shannon contributed to the field of cryptanalysis for the national defence during World War II, including his fundamental work on code breaking and securing telecommunications (Fig. 3.4).

**George Stibitz's Computer:** George Robert Stibitz (1904–1995) was a Bell Labs researcher, internationally recognized as one of the fathers of the first modern digital computer (Fig. 3.5).

At the end of 1938, Bell Labs launched the research computer project led by Stibitz. Created under his leadership, the computer was completed on January 8, 1940, and was able to perform operations with complex numbers. On September 11, 1940, Stibitz had demonstrated his computer at the American Mathematical Society conference at Dartmouth College. This was the first attempt to use a remote computer by using a telephone line. Among the conference participants, who witnessed the demonstration, were John

Fig. 3.5. George Robert Stibitz (1904–1995).

von Neumann, John Mockley and Norbert Wiener, who later wrote about this demonstration in their memoirs.

**John von Neumann Principles:** Leibniz's brilliant predictions about the role of the *binary system* in modern computer science came true only after 2,5 centuries, when the eminent American scientist, physicist and mathematician John von Neumann (1903–1957), after a thorough analysis of the advantages and disadvantages of the first electronic computer in the computer history, ENIAC, gave a decisive preference for the *binary system* as the *universal way of encoding–decoding information in electronic computers*. Regardless of Neumann, the eminent Soviet scientist academician Sergey Lebedev (1902–1974), article at the general designer of the first Soviet computers, arrived at the same principles. Therefore, in the Russian historical and mathematical literature, the fundamental principles of the construction of the modern electronic computers are also called *Neumann–Lebedev principles* (Figs. 3.6 and 3.7).

Neumann's idea to use the *binary system* in the *electronic digital computers* had been perfect due to its *indisputable arithmetical advantages* for the case of the *binary electronic elements*. On this occasion, Neumann wrote as follows:

> *"Our main memory block is adapted by its nature to the binary system ... The trigger in essence is also the binary device ... But the main advantage of the binary system compared to the decimal*

Fig. 3.6.  John von Neumann (1903–1957).

Fig. 3.7.  Sergey Lebedev (1902–1974).

*one is greater simplicity and higher speed, with which elementary operations can be performed …*

*An additional remark is that the main part of the machine is not arithmetical, but logical. The new logic, being of **"yes-no"** logic, is mostly **binary**. Therefore, the binary construction of arithmetic devices will significantly promote to the creation of the more homogeneous machine that can be better configured and more efficient."*

## 3.2. Ternary Symmetric Numeral System and Computer "Setun"

### 3.2.1. The concept of canonical positional numeral system

The *Babylonian numeral system* with base 60, the *decimal* and *binary* numeral systems, considered above, belong to the class of the so-called *canonical* positional numeral systems [135], which is given by the following general formula:

$$x = \sum_{i=-k}^{n} b_i R^i, \qquad (3.1)$$

where $x$ is a real number (integer, fractional or mixed), $R$ is the base of the numeral system (3.1), $b_i$ is the numeral, and $R^i$ is the weight of the $i$th digit. The numerals of the system (3.1) take the values from the alphabet $A = \{a_1, a_2, \ldots, a_m\}$. For the decimal system, the alphabet $A$ consists of 10 numerals, i.e., $A = \{0, 1, 2, 3, 4, 5, 6, 7, 8, 9\}$.

The abbreviated notation of the system (3.1) is as follows:

$$x = b_n b_{n-1} \ldots b_0, b_{-1} b_{-2} \ldots b_{-k}. \qquad (3.2)$$

It is generally accepted in the canonical system (3.1) that one can use some integers greater than or equal to 2 as base $R$. In this case, the peculiarity of the *canonical* system (3.1) is that its base $R$ and the number of numerals $m$ of the alphabet $A = \{a_1, a_2, \ldots, a_m\}$ coincide, i.e., $R = m$.

In addition to the *decimal*, other positional numeral systems of the type (3.1), having the bases $2, 3, 8, 12, 16, 60$, are also widely used in mathematics and computer science:

**2** is the *binary system* (in discrete mathematics, computer science, programming);

**3** is the *ternary system* (in computer science);

**8** is the *octal system* (in programming);

**12** is the *12-decimal system* (calculation by dozens);

**16** is the *hexadecimal system* (in programming and computer science);

**60** is the *Babylonian system* with the base of 60 (measurement of time and angles, in particular, coordinates, longitude and latitude);

### 3.2.2. The concept of the "symmetric" numeral system

The "symmetric" numeral systems are a generalization of the notion of the canonical numeral system (3.1), while this generalization is carried out by expanding the concept of the *numeral* in the canonical system (3.1).

As well known [135], the numerical equivalents of the *numerals* in the canonical system (3.1), as a rule, are *positive integers*, which take the values from the set $\{2, \ldots, R - 1\}$, where $R$ is the base of the numeral system. The main disadvantage of the canonical numeral system (3.1) is the *impossibility of representing negative numbers* without introducing a special *sign digit*. As it is known, the concepts of the *inverse* and *additional* codes are widely used to represent the *negative numbers* in the digital computers. Thus, the *negative numbers* within the framework of the traditional canonical systems (3.1) play the role of *outcasts*, which cannot be represented in the traditional canonical system in the *direct code*. To eliminate this drawback, the so-called *symmetric* positional numeral systems were introduced [135].

Let's consider the following "*symmetric*" *positional numeral system*:

$$N = \sum_{i=0}^{n-1} b_i R^i, \tag{3.3}$$

where $R = 2S + 1$, $b_i \in \{\overline{S}, \overline{S-1}, \ldots, \overline{1}, 0, 1, 2, \ldots, S\}$, $\overline{S} = -S$.

Thus, the *symmetric system* (3.3) has two main features:

(1) The base of the *symmetric system* (3.3) is always the *odd positive integer* $R = 2S + 1$, that is, $R = 3, 5, 7, \ldots$.
(2) The "numerals" of the *symmetric system* (3.3) take the values from the set $b_i \in \{\overline{S}, \overline{S-1}, \ldots, \overline{1}, 0, 1, 2, \ldots, S\}$, where $\overline{S} = -S$. This means that the *symmetric system* (3.3) uses $2S + 1$

*numerals*: one of these *numerals* has the numerical equivalent of 0; the remaining $2S$ *numerals* are divided into two groups: the *numerals of the first group have positive numerical equivalents* $1, 2, 3, \ldots, S$; the *numerals of the second group have the negative equivalents* $\overline{1}, \overline{2}, \overline{3}, \ldots, \overline{S}$. Such a choice of the set of *numerals* has significant advantages in comparison to the *classical canonical systems* (3.1): *In the "symmetrical" system* (3.3), *the negative numbers can be represented without introducing the concepts of the inverse and additional codes.*

It has been proved [135] that in the *symmetric system* (3.3), we can represent the $R^n$ integers (positive and negative) in the range from

$$N_{\min} = -\frac{R^n - 1}{2} \quad \text{to} \quad N_{\max} = \frac{R^n - 1}{2}. \tag{3.4}$$

### 3.2.3. Ternary symmetric numeral system

As well known, the binary system, which uses the two *numerals* $\{0, 1\}$, is the simplest among the canonical systems (3.1); among the symmetric numeral systems of the type (3.3), the simplest is the *ternary symmetric system*, which has the following numerical parameters: *the base of the system equals* $R = 3$ *and the ternary symmetric system used three ternary numerals* $\{\overline{1}, 0, 1\}$.

For the case of $R = 3$, the representation of integers in the ternary "symmetric" system (3.3) has the following form:

$$N = \sum_{i=0}^{n-1} b_i 3^i, \tag{3.5}$$

where $b_i \in \{\overline{1}, 0, 1\}$ is the *ternary numeral* and $3^i$ is the weight of the *i*th digit.

It is clear that the abbreviated records of the *maximum* and *minimum* numbers in the *ternary "symmetric"* system (3.5) have the following form, respectively:

$$N_{\max} = \underbrace{111 \ldots 1}_{n}, \tag{3.6}$$

$$N_{\min} = \underbrace{\overline{1}\,\overline{1}\,\overline{1} \ldots \overline{1}}_{n}. \tag{3.7}$$

The digital records (3.6) and (3.7) have the following algebraic interpretation:

$$N_{\max} = 3^{n-1} + 3^{n-2} + \cdots + 3^1 + 3^0 = \frac{3^n - 1}{2}, \tag{3.8}$$

$$N_{\min} = -3^{n-1} - 3^{n-2} - \cdots - 3^1 - 3^0 = -\frac{3^n - 1}{2}. \tag{3.9}$$

Then it follows from the above reasoning that in the *ternary symmetric numeral system* (3.5), by using the $n$ digits, we can represent the $3^n$ integers (including positive, negative numbers and 0) in the range from $N_{\min} = -\frac{3^n-1}{2}$ to $N_{\max} = \frac{3^n-1}{2}$.

### 3.2.4. Representation of the negative numbers

Let's consider the two $n$-digit numbers $A$ and $B$, represented in the *ternary "symmetric" system* (3.5), as follows:

$$A = 1b_{n-2}b_{n-3}\ldots b_1 b_0, \tag{3.10}$$

$$B = \bar{1}b_{n-2}b_{n-3}\ldots b_1 b_0. \tag{3.11}$$

Note that number $A$ contains the positive unit of 1, and number $B$ contains the negative unit of $\bar{1}$ in the most significant digit of the digital ternary "symmetric" representations (3.10) and (3.11), wherein the *ternary numerals* $b_{n-1}, b_{n-2}, \ldots, b_1, b_0$ from the representations (3.10) and (3.11) may take some values from the set $\{\bar{1}, 0, 1\}$.

It is easy to prove that, whatever be the values of the *ternary numerals* $b_{n-1}, b_{n-2}, \ldots, b_1, b_0$, the digital record of the number (3.10) always represents only the positive number, regardless of the values of the remaining digits $b_{n-1}, b_{n-2}, \ldots, b_1, b_0$ in the digit record (3.10). This conclusion follows directly from the inequality:

$$3^{n-1} > \frac{3^{n-1} - 1}{2}, \tag{3.12}$$

where $3^{n-1}$ is the weight of the most significant, that is, the $(n-1)$th digit in the ternary digital record (3.10), and $\frac{3^{n-1}-1}{2}$ is the sum of the weights of the remaining digits $b_{n-1}, b_{n-2}, \ldots, b_1, b_0$ in the ternary digital record (3.10).

For the case (3.11), the inequality (3.12) turns into another inequality:

$$-3^{n-1} < -\frac{3^{n-1} - 1}{2}. \qquad (3.13)$$

It follows from these arguments that the ternary "symmetric" representation (3.11) always represents only the negative integer, regardless of the values of the remaining digits, $b_{n-1}, b_{n-2}, \ldots, b_1, b_0$ in the ternary "symmetric" representation (3.11).

From these arguments, there follows a very important conclusion that the information about the sign of the ternary number, which is represented in the *ternary symmetric system* (3.5), is contained in the most significant digit of the ternary number; in this case, if the most significant digit of the ternary number is the positive unit $(+1)$, then the ternary number is positive, if the most significant digit is the negative unit $(\bar{1})$, then the ternary number is negative.

Thus, we have found a very important property of the ternary "symmetric" numeral system (3.5). This system does not have a special symbol to represent the sign of the ternary number. Positive and negative numbers in this system are represented in the *direct* code. Moreover, this property is invariant with respect to arithmetical operations. This means that all arithmetical operations in the *ternary symmetric system* (3.5) can be performed in the *direct* code without using the concepts of the *inverse* and *additional* codes.

It should be noted that an important advantage of the *ternary symmetric system* (3.5) is the existence of a very simple rule for obtaining the ternary number of the opposite sign from the original ternary number. For this, the rule of the *ternary inversion* is applied to the initial *ternary symmetric representation*, the essence of which is as follows:

$$\bar{1} \to 1; \; 0 \to 0; \; 1 \to \bar{1}. \qquad (3.14)$$

**Example 7.1.** Let the initial representation of the decimal number $56_{10}$ in the *ternary "symmetric"* system (3.5) be in the following

Table 3.1. The rule of the summation of the single-digit ternary "symmetric" numbers $a_k + b_k$.

| $a_k/b_k$ | $\bar{1}$ | 0 | 1 |
|-----------|-----------|---|---|
| $\bar{1}$ | $\bar{1}\,1$ | $\bar{1}$ | 0 |
| 0 | $\bar{1}$ | 0 | 1 |
| 1 | 0 | 1 | $1\,\bar{1}$ |

form:

$$56_{10} = 1\bar{1}01\bar{1}_3.$$

Then, from this *ternary symmetric representation*, by using the rule (3.14), it is easy to get the *ternary symmetric representation* of the opposite sign, that is,

$$-56_{10} = \bar{1}\,1\,0\bar{1}\,1_3.$$

### 3.2.5. Ternary "symmetric" arithmetic

The basis of *the ternary symmetric summation* is the following elementary identity, which connects the powers of the number 3:

$$2 \times 3^i = 3^{i+1} - 3^i. \tag{3.15}$$

Table 3.1, which sets forth the rule for the summation of the *ternary symmetric single-digit numbers*, follows from the identity (3.15).

A very simple rule of the *ternary symmetric summation* of the *single-digit ternary symmetric numbers* $a_k + b_k$ follows from Table 3.1. The features of such a summation arise when we summarize the ternary "symmetric" single-digit units of one and the same sign:

$$1 + 1 = 1\,\bar{1} \quad \text{and} \quad \bar{1} + \bar{1} = \bar{1}\,1. \tag{3.16}$$

It follows from the expressions (3.16) that at the summation of the ternary "symmetric" single-digit units of the same sign, there arise the *carryover to the higher digit* and the *intermediate sum* in the same digit. In this case, the *carryover* has the same sign as the sign of the summable units and the intermediate sum of the opposite sign in the same digit.

Table 3.2. The multiplication rule of the single-digit ternary "symmetric" numbers $a_k \times b_k$.

| $a_k/b_k$ | $\bar{1}$ | 0 | 1 |
|-----------|-----------|---|---|
| $\bar{1}$ | 1 | 0 | $\bar{1}$ |
| 0 | 0 | 0 | 0 |
| 1 | $\bar{1}$ | 0 | 1 |

The *ternary symmetric summation* is the main arithmetic operation in the above considered numeral system (see Table 3.1). Indeed, it is easy to show that the *ternary symmetric subtraction* in this system reduces to the summation. Suppose we need to perform the *ternary symmetric subtraction* of the two ternary "symmetric single-unit numbers: $A - B$. Let's represent the difference $A - B$ in the following form:

$$A - B = A + (-B). \tag{3.17}$$

As it follows from (3.17), the subtraction $A - B$ reduces to the summation if we apply to the subtrahend $B$ the rule of the ternary inversion (3.14).

The *ternary symmetric multiplication* is based on the following trivial identity, which links the powers of number 3:

$$3^m \times 3^n = 3^{m+n}. \tag{3.18}$$

Table 3.2, which gives the rule of the multiplication of the ternary "symmetric" single-digit numbers $a_k \times b_k$, follows from (3.18).

The ternary symmetric division is similar to the division in the classical binary system and reduces to the shift of the divider and its subtraction from the dividend.

The following examples demonstrate the rules of the ternary symmetric summation and multiplication:

**Multiplication**

**Summation**

$$
\begin{array}{r}
0\;1\;1\;1\;0\;\bar{1} \\
0\;0\;1\;\bar{1}\;\bar{1}\;1 \\
\hline
1\;\bar{1}\;\bar{1}\;0\;\bar{1}\;0
\end{array}
$$

$$
\begin{array}{r}
\bar{1}\;\;0\;\;\bar{1} \\
\times \quad 1\;\;\bar{1} \\
\hline
1\;\;0\;\;1 \\
\bar{1}\;\;0\;\;\bar{1} \quad\;\; \\
\hline
\bar{1}\;\;1\;\;\bar{1}\;\;1
\end{array}
$$

Fig. 3.8. Nikolay Brusentsov (1925–2014).

### 3.2.6. Ternary computer "Setun"

The *ternary symmetric numeral system* (3.1) is the key idea of the new principle of computer design [135]. This principle is based on the following mathematical ideas:

(1) *ternary logic*;
(2) *ternary memory element*;
(3) *ternary symmetric numeral system* (3.3).

By using this principle, called the *Brousentsov ternary principle* in honor of its creator Nikolay Brusentsov (1925–2014) [72], the computer "Setun" was created at the dawn of the computer era (Fig. 3.8). This ternary computer was designed on magnetic elements, and therefore, it wasn't concurrent to the binary electronic computers based on the electronic elements. However, its architecture, based on the *Brousentsov ternary principle*, turned out to be so perfect that at present, the project involving this computer attracts great attention from many computer experts.

The first sample of the ternary computer "Setun" was designed at Moscow University in 1958. Computer adjustment was carried out in record time. The computer had earned already on the 10th day of the complex adjustment.

The serial mastering of the computer "Setun" was carried out by the Kazan Factory of Mathematical Machines in the 1960s. There were 50 copies of this machine, which worked reliably and productively in all climatic zones from Kaliningrad to Magadan, from Odessa and Ashgabat to Novosibirsk and Yakutsk.

The famous Soviet computer specialist Professor Dmitry Pospelov appreciated this unique computer project as follows [135]:

> *"The barriers that stand on the way of the application of the ternary symmetric numeral system in computer technology have obstacles of technical characters. Until now, the economical and efficient elements with three stable states hadn't been designed. As soon as such elements will be designed, the most of the computers of the universal type and many specialized computers are likely to be designed so that they will be functioned in the ternary symmetric numeral system."*

Also, the prominent American scientist Donald Knuth expressed the opinion that the replacement of the *binary trigger* (*flip-flop*) with the *ternary trigger* (*flip-flap-flop*) will happen one day.

The *binary system* and the *ternary symmetric numeral systems* are the most prominent examples of how numeral systems can affect the development of information technology. The *binary system* has already led to the origin of *modern binary informational technology*, based on the *Neumann Binary Principles*. There is every reason to believe that the *ternary symmetric numeral system* (3.1) can lead to a new *ternary informational technology* or even to the *ternary information revolution* because this scientific direction is rapidly developing in modern informatics.

Unfortunately, like many original Soviet developments, the fate of the ternary computer "Setun" was tragic. The leadership of Moscow University could not assess the revolutionary importance of Nikolay Brusentsov's computer projects in modern computer science.

In his wonderful book *The History of Computer Engineering in Persons*, Kiev, Firm "KIT", 1995, the corresponding member of the Ukrainian Academy of Sciences Boris Malinovsky describes this tragedy as follows: *"The designing of computers is not a matter of* the University science, such the leadership of the Moscow University decided. The first model of the Brousentsov's Setun machine (the experimental model that worked 17 years without failures) was barbarously destroyed; it was cut into pieces and thrown into a landfill."

# Chapter 4

# Bergman's System and "Golden" Number Theory

## 4.1. George Bergman and Bergman's System

### 4.1.1. Definition of Bergman's system

In 1957, the young American mathematician George Bergman published the article "A number system with an irrational base" in the authoritative journal, *Mathematics Magazine* [54]. The following sum is called *Bergman's system*:

$$A = \sum_i a_i \Phi^i, \qquad (4.1)$$

where $A$ is any real number, $a_i$ is the binary numeral $\{0,1\}$ of the $i$th digit, $i = 0, \pm1, \pm2, \pm3, \ldots,$ $\Phi^I$ is the weight of the $i$th digit, and $\Phi = (1 + \sqrt{5})/2$ is the base of the numeral system (4.1).

### 4.1.2. George Bergman

Detailed information about George Bergman can be obtained from Wikipedia Fig. 4.1.

> *"George Mark Bergman was born on 22 July 1943 in Brooklyn, New York. He received his PhD from Harvard in 1968, under the direction of John Tate. The year before, he had been appointed Assistant Professor of mathematics at the University of California, Berkeley, where he has taught ever since, being promoted to Associate Professor in 1974 and to Professor in 1978. His primary*

Fig. 4.1. George Mark Bergman.

*research area is algebra, in particular associative rings, universal algebra, category theory and the construction of counterexamples. Mathematical logic is an additional research area. Bergman officially retired in 2009, but is still teaching."*

It is interesting to note that the concept of Bergman's system has entered widely into Internet and modern scientific literature. The special article in Wikipedia is dedicated to *Bergman's system.* It is described briefly in Wolfram MathWorld. Donald Knuth refers to Bergman's article [54] in his outstanding book [123]. The special paragraph in Stakhov's book [6] is dedicated to Bergman's system. *The Computer Journal* (British Computer Society) had in 2002 published Stakhov's article [72]; this article is based on Bergman's system (4.1) and is dedicated to the so-called *ternary mirror-symmetrical arithmetic,* which was evaluated highly by Prof. Donald Knuth. Thus, the article [54] had glorified Bergman's name more than his other mathematical works published during his adulthood. It is surprising that there is no mention about Bergman's outstanding article [54] in Wikipedia.

### 4.1.3. The main distinction between Bergman's system and binary system

On the face of it, there is no essential distinction between the formula (4.1) for Bergman's system and the formulas for the canonical positional numeral systems, in particular, the *binary system*:

$$A = \sum_i a_i 2^i (i = 0, \pm1, \pm2, \pm3, \ldots) \quad (a_i \in \{0,1\}), \qquad (4.2)$$

where the digit weights are connected by the following "arithmetical" relations:

$$2^i = 2^{i-1} + 2^{i-1} = 2 \times 2^{i-1}, \qquad (4.3)$$

which underlie "binary arithmetic".

However, the principal distinction between the *Bergman system* (4.1) and the *binary system* (4.2) is the fact that the irrational number $\Phi = (1 + \sqrt{5})/2$ (the golden ratio) is used as the base of the numeral system (4.1) and its digit weights are connected by the following relations:

$$\Phi^i = \Phi^{i-1} + \Phi^{i-2} = \Phi \times \Phi^{i-1}, \qquad (4.4)$$

which underlie the *"golden" arithmetic*.

This is why Bergman called his mathematical result the *numeral system with irrational base*. Although Bergman's article [54] provides the fundamental result of the number theory and computer science, mathematicians and experts of computer science during that period were not able to appreciate the mathematical discovery of the American wunderkind.

## 4.2. The "Golden" Number Theory and the New Properties of Natural Numbers

### 4.2.1. Euclid's definition of natural numbers

It is well known that a *number* is one of the most important notions of mathematics and the *number theory* is one of the most famous ancient mathematical theories.

But what is the *number*? It would seem at first sight that mathematicians arrived at a common answer to this question. But all was not so simple. There are various definitions of the *number*. The simplest of them is well known as the definition of *natural numbers* used by Euclid in his *Elements*.

Euclid developed a geometric approach to the notion of the *number*. He treats all the *numbers* as *geometric segments*, and such geometric approach led him to the following geometric definition of *natural numbers*. Suppose that we have the infinite number of *standard segments* of the unit length of 1. Euclid named them the *monads* and he did not consider the *monads* as *numbers*. It was simply the *beginning of all numbers*. It is clear that for the construction of all *natural numbers*, we should have the infinite set $S$ of the *monads*, that is,

$$S = \{1, 1, 1, \ldots\}. \tag{4.5}$$

Then we can define the *natural number* $N$ as some *geometric segment*, which can be represented as the *sum of the monads*, taken from (4.5), that is,

$$N = \underbrace{1 + 1 + 1 + \cdots + 1}_{N}. \tag{4.6}$$

In spite of the limiting simplicity of the definition (4.6), it had played a great role in mathematics, in particular, in the *number theory*. This definition underlies many important mathematical concepts, for example, concepts of the *prime* and the *composed* numbers, the definition (4.6) also underlies the concept of *divisibility*, which is one of the main concepts of the *number theory*.

### 4.2.2.  Constructive definition of the real numbers

But there are also other definitions of the *number*. For example, the so-called *constructive approach* to the definition of the *real number* is known. According to this approach, the *real number* $A$ is some mathematical object, which can be represented in the *binary system* (4.2). The definition of the real number $A$, given by (4.2), has the following geometric interpretation. Let's now consider the infinite set of the *binary segments* of the length $2^n$, that is,

$$B = \{2^n\} \quad (n = 0, \pm 1, \pm 2, \pm 3, \ldots). \tag{4.7}$$

Then, all the real numbers can be represented by the sum (4.2), which consists of the *binary segments* taken from (4.7).

Note that the number of terms, included in the sum (4.2), is always *finite* but *potentially unlimited*, that is, the definition (4.2) is a brilliant example of the *potential infinity concept* used in *constructive mathematics* [124].

Clearly, the definition (4.2) sets on the numerical axis only a part of the real numbers, which can be represented by the *finite* sum (4.2). We will name such numbers *constructive real numbers*. All other real numbers, which cannot be represented by the *finite* sum (4.2), are *non-constructive real numbers*.

What numbers can be referred to as *non-constructive numbers* within the framework of the definition (4.2)? Clearly, all the *irrational numbers*, in particular, the mathematical constants $\pi$ and $e$, the number $\sqrt{2}$, and the *golden ratio* are referred to as *non-constructive numbers*. But within the framework of the definition (4.2), some *rational numbers* (for example, $2/3$, $3/7$, etc.), which cannot be represented by the *finite* sum (4.2), should be referred to as *non-constructive numbers*.

Note, that though the definition (4.2) considerably limits the set of real numbers, this fact does not belittle its significance from the "practical", computing point of view. It is easy to prove that any *non-constructive real number* can be represented in the form (4.2) approximately, and the approximation error $\Delta$ will decrease with increasing terms in (4.2), however $\Delta \neq 0$ for all the *non-constructive real numbers*. In essence, in modern computers, we only use the *constructive numbers*, given by (4.2), however we do not have any problem with the *non-constructive numbers* because they can be represented in the form (4.2) with the approximation error $\Delta \neq 0$, which strives to be 0 potentially.

### 4.2.3. Newton's definition of the real number

During many millennia, the mathematicians developed and précised the concept of the *number*. In the 17th century during the origin of modern science, in particular, modern mathematics, a number of methods of studying the *continuous processes* were developed and the concept of the *real number* comes to the foreground. Most clearly, the new definition of this concept was given by Isaac Newton, one of the founders of the *mathematical analysis* in his *Arithmetica Universalis* (1707):

> *"We understand a number not as the set of units, but as the abstract ratio of some magnitude to other one of the same kind, taken as the unit."*

This formulation gives the *universal definition* of the *real numbers, rational* or *irrational*. If you now consider *Euclid's definition* (4.6) from the point of *Newton's definition*, we can see that the *monad* in (4.6) plays the role of the *unit*. In the *binary system* (4.2), the number 2, that is, the base of the *binary system*, plays the role of the *unit*.

### 4.2.4.  The "extended" Fibonacci and Lucas numbers

*Bergman's* system (4.1) is closely connected with the so-called *extended Fibonacci and Lucas numbers* $F_i$ and $L_i$ ($i = 0, \pm 1, \pm 2, \pm 3, \ldots$) introduced in Chapter 1 of Volume I (see Table 4.1).

As shown in Chapter 1, the "extended" Fibonacci and Lucas numbers are connected by the following relations:

$$F_{-n} = (-1)^{n+1} F_n; \quad L_{-n} = (-1)^n L_n. \qquad (4.8)$$

### 4.2.5.  Bergman's system as a new definition of the real number

In the previous sections, we developed the so-called *constructive approach* to the definition of the *real numbers*, based on the *binary*

Table 4.1. The "extended" Fibonacci and Lucas numbers.

| $n$ | 0 | 1 | 2 | 3 | 4 | 5 | 6 | 7 | 8 | 9 | 10 |
|---|---|---|---|---|---|---|---|---|---|---|---|
| $F_n$ | 0 | 1 | 1 | 2 | 3 | 5 | 8 | 13 | 21 | 34 | 55 |
| $F_{-n}$ | 0 | 1 | −1 | 2 | −3 | 5 | −8 | 13 | −21 | 34 | −55 |
| $L_n$ | 2 | 1 | 3 | 4 | 7 | 11 | 18 | 29 | 47 | 76 | 123 |
| $L_{-n}$ | 2 | −1 | 3 | −4 | 7 | −11 | 18 | −29 | 47 | −76 | 123 |

*system* (4.2), and this idea allows the following generalization. We can extend Newton's definition of *real numbers* for the case of *Bergman's system* (4.1). In fact, such interpretation of *Bergman's system* (4.1) has great theoretical importance for modern mathematics and its history. It turns over our ideas about the *real numbers.* Historically, *natural numbers* were the first class of *real numbers*; the *irrational numbers* were introduced into mathematics much later after the discovery of the *incommensurable segments.* In the traditional numeral systems (Babylonian sexagesimal, decimal, binary), some natural numbers, 60, 10, 2, were used as the *beginning of calculus.* All the *real numbers* can be represented with the help of *bases* 60, 10, or 2. In *Bergman's system* (4.1), the golden ratio $\Phi = (1 + \sqrt{5})/2$ is the *beginning of calculus.* All the other real numbers (including natural numbers) can be represented through the golden ratio $\Phi = (1 + \sqrt{5})/2$. This means that the irrational number $\Phi = (1 + \sqrt{5})/2$ (the golden ratio) is becoming the *major number of mathematics* because it allows the representation of all numbers (natural, rational, irrational) in the *Bergman system* (4.1), as shown in the following. But this conclusion fully coincides with the *harmonic ideas* of Pythagoras, Plato, Euclid, and in later periods (Renaissance and the 19th century) with the ideas of Kepler, Pacioli, Klein, Binet, Lucas, and in the 20th and 21st centuries with the ideas of prominent mathematicians and philosophers, including Turing, Hoggatt, Coxeter, Vorobyov and others.

### 4.2.6. The "golden" representation of natural numbers

A new definition of the real numbers, based on *Bergman's system* (4.1), can become a source for the *new number-theoretical results.* We begin our research from the *"golden" representations of natural numbers* in *Bergman's system* (4.1). With this purpose, we will study the following representation of the *natural numbers N* in the *Bergman system* (4.1):

$$N = \sum_i a_i \Phi^i, \tag{4.9}$$

where $a_i \in \{0, 1\}$ is the bit and $\Phi^i$ is the weight of the $i$th digit, $\Phi = \frac{1+\sqrt{5}}{2}$ is the base of the numeral system (4.9).

We will name the sum (4.9) as the $\Phi$-*code of the natural number* $N$. The abridged notation of the $\Phi$-code of the natural number $N$ has the following form:

$$N = a_n a_{n-1} \ldots a_1 a_0 a_{-1} a_{-2} \ldots a_{-k} \qquad (4.10)$$

and is named the "*golden*" *representation of the natural number* $N$.

Note that the point in the expression (4.10) separates the "*golden*" *representation* (4.10) into two parts: the left-hand part, where the bits $a_n a_{n-1} \ldots a_1 a_0$ have the *non-negative* indices, and the right-hand part, where the bits $a_{-1} a_{-2} \ldots a_{-k}$ have the *negative* indices.

Note that the weights $\Phi^i$ of the $\Phi$-*code* (4.9) are connected by the following relations:

$$\Phi^i = \Phi^{i-1} + \Phi^{i-2}, \qquad (4.11)$$

$$\Phi^i = \Phi \times \Phi^{i-1}. \qquad (4.12)$$

The relations (4.11) and (4.12) are called *additive* and *multiplicative* relations for Bergman's system, respectively.

Besides, the power of the golden ratio $\Phi^i$ is expressed through the "extended" Fibonacci and Lucas numbers (see Table 4.1) as follows:

$$\Phi^i = \frac{L_i + F_i \sqrt{5}}{2} \quad (i = 0, \pm 1, \pm 2, \pm 3, \ldots). \qquad (4.13)$$

Note that the formula (4.13) is a generalization of the formula $\Phi = \frac{1+\sqrt{5}}{2}$ (for the *golden ratio*) because for the case $i = 1$, we have $L_1 = F_1 = 1$.

Let's now apply the rule

$$N' = N + 1 \qquad (4.14)$$

for obtaining all the natural numbers.

In order to apply the rule (4.14) for obtaining all of the "golden" representations of the natural numbers in the form (4.10), we need to transform the "golden" representation (4.10) of the initial number

$N$ to such form, when the bit of the zeroth digit will become equal to 0, i.e., $a_0 = 0$. We can fulfill such transformation by means of the micro-operations of the *convolutions* and *devolutions* based on the *additive* relation (4.11):

$$\text{Convolution}: \boxed{100} \leftarrow \boxed{011}, \tag{4.15}$$

$$\text{Devolution}: \boxed{100} \rightarrow \boxed{011}. \tag{4.16}$$

If we add the binary 1 to the zeroth digit of the *"golden"* *representation* (4.10), we carry out the rule (4.14).

We begin the demonstration of this method from the transformation of the *"golden"* *representation* of number 1 to the "golden" representation of number 2. The *"golden"* *representation* of number 1 can be represented as follows:

$$1 = 1.00 = 1 \times \Phi^0 + 0 \times \Phi^{-1} = \Phi^0. \tag{4.17}$$

By using the micro-operation of the *devolution* (4.16), we get other *"golden"* *representation* $\boxed{0.11}$ of number 1 as follows:

$$1 = \boxed{1.00} = \boxed{0.11} = \Phi^{-1} + \Phi^{-2} = 0.618 + 0.382. \tag{4.18}$$

After this, we can apply the rule (4.14) to the *"golden"* *representation* $\boxed{0.11}$ in (4.18), that is, we add the bit of 1 to the zeroth digit of the "golden" representation $\boxed{0.11}$ in (4.18). As a result, we get the *"golden"* *representation* of number 2:

$$2 = 001.11 = \Phi^0 + \Phi^{-1} + \Phi^{-2} = 1 + 0.618 + 0.382. \tag{4.19}$$

If we apply the micro-operation of the *convolution* (4.15) to the *"golden"* *representation* (4.19), we get the other *"golden"* *representation* of number 2:

$$2 = 0\boxed{01.1}1 = 0\boxed{10.0}1 = \Phi^1 + \Phi^{-2} = 1.618 + 0.382. \tag{4.20}$$

By adding the bit of 1 to the zeroth digit of the *"golden"* *representation* (4.20) and by carrying out the *convolution* (4.15), we get the following *"golden" representation* of number 3:

$$3 = 0\boxed{011}.01 = 0\boxed{100}.01 = \Phi^2 + \Phi^{-2} = 2.618 + 0.382. \qquad (4.21)$$

The *"golden" representation* of number 4 follows from (4.21):

$$4 = 0101.01 = \Phi^2 + \Phi^0 + \Phi^{-2} = 2.618 + 1 + 0.382. \qquad (4.22)$$

We get the *"golden" representation* of number 5 from the *"golden" representation* (4.22) if we carry out the following transformation in (4.22) by using the *devolutions*:

$$4 = 101.0\boxed{100} = 101.0\boxed{011} = 10\boxed{1.00}11 = 10\boxed{0.11}11 = 100.1111$$

$$= \Phi^2 + \Phi^{-1} + \Phi^{-2} + \Phi^{-3} + \Phi^{-4}$$

$$= 2.618 + 0.618 + 0.382 + 0.236 + 0.146. \qquad (4.23)$$

By adding the bit of 1 to the zeroth digit of the right-hand *"golden" representation* (4.23) (100.1111), we get the following "golden" representation of number 5:

$$5 = 100.1111 = 101.1111 = 2.618 + 1 + 0.618$$

$$+ 0.382 + 0.236 + 0.146. \qquad (4.24)$$

By continuing this process *ad infinitum*, we can get the *"golden" representations* of all the natural numbers. Thus, this consideration leads us to an unexpected mathematical result, which can be formulated in the form of the following theorem.

**Theorem 4.1.** *All natural numbers can be represented in the $\Phi$-code (4.9) of Bergman's system (4.1) by the finite number of bits.*

We note that Theorem 4.1 is far from trivial if we take into consideration that all the powers of the *golden proportion* $\Phi^i (i = \pm 1, \pm 2, \pm 3, \ldots)$ in (4.9) (with the exception of $\Phi^0 = 1$) are irrational numbers.

Note that Theorem 4.1 is true only for the natural numbers. Therefore, Theorem 4.1 can be referred to as the category of the *new properties of natural numbers.*

### 4.2.7. Multiplicity and MINIMAL FORM of the "golden" representations

As it follows from the above examples (4.18)–(4.24), the main feature of the *"golden" representations* of the real numbers in Bergman's system (4.1), compared to the *binary system* (4.2), is the *multiplicity* of the *"golden" representations* of the same real number. The various *"golden" representations* of one and the same real number can be obtained by using the micro-operations of *convolutions* (4.15) and *devolutions* (4.16) over the "golden" representations (4.10).

A special role among the various *"golden" representations* (4.10) of one and the same number is played by the so-called MINIMAL FORM (MΦ), which can be obtained from the initial "golden" representation by carrying out all the possible *convolutions* (4.15) over the *"golden" representation* (4.10).

The MINIMAL FORM has the following important features:

(1) The micro-operation of the *convolution* ($\boxed{100} \leftarrow \boxed{011}$) reduces to the transformation of the triple of the neighbouring bits 011 into another triple of the neighbouring bits 100; this means that *in the MINIMAL FORM, two bits of 1 do not meet alongside*.

(2) The MINIMAL FORM has the minimal number of 1's among all the possible "golden" representations of one and the same integer $N$.

### 4.3. $Z$- and $D$-properties of Natural Numbers

Bergman's system (4.1) is a source for the new number-theoretical results. The $Z$-property of natural numbers is one of them. This property follows from the next very simple reasoning.

Let's consider the Φ-*code* (4.9). Note that according to Theorem 4.1, the sum (4.9) is the *finite sum* for the arbitrary natural number $N$.

If we substitute the formula (4.11) instead of $\Phi^i$ in the formula (4.9), we can represent the Φ-*code* (4.9) in the following form:

$$N = \frac{A + B\sqrt{5}}{2}, \tag{4.25}$$

where

$$A = \sum_i a_i L_i, \qquad (4.26)$$

$$B = \sum_i a_i F_i. \qquad (4.27)$$

Note that all the binary numerals in the sums (4.26) and (4.27) coincide with the corresponding bits of the $\Phi$-*code* (4.9).

Let's now represent the expression (4.25) in the following form:

$$2N = A + B\sqrt{5}. \qquad (4.28)$$

Note that the expression (4.28) has a general character and is valid for the arbitrary natural number $N$.

Let's analyze the expression (4.28). It is clear that the number $2N$, standing in the left-hand part of the expression (4.28), is always the *even* number. The right-hand part of the expression (4.28) is the sum of the number $A$ and the product of the number $B$ by the irrational number $\sqrt{5}$. But according to (4.26) and (4.27), the numbers $A$ and $B$ are always integers because the Fibonacci and Lucas numbers are integers. Then it follows from (4.28) that for the given natural number $N$, the *even* number $2N$ is identically equal to the sum of the integer $A$ and the product of the integer $B$ on the irrational number $\sqrt{5}$. This statement is valid for the arbitrary natural number $N$!

Then, there is the question: For those conditions, is the identity (4.28) valid for the arbitrary natural number? The answer to this question is very simple: the identity (4.28) can be valid for the arbitrary natural number $N$ only for the case where the sum (4.27) is identically equal to 0 ("zero") and the sum (4.26) is identically equal to the double number of $N$, that is,

$$B = \sum_i a_i F_i \equiv 0, \qquad (4.29)$$

$$A = \sum_i a_i L_i \equiv 2N. \qquad (4.30)$$

Let's now compare the sums (4.29) and (4.9). As the bits $a_i(i = 0, \pm 1, \pm 2, \pm 3, \ldots)$ in these sums coincide, the expression (4.29) can be obtained from the expression (4.9) if we replace every power of the golden ratio $\Phi^i(i = 0, \pm 1, \pm 2, \pm 3, \ldots)$ in the $\Phi$-code (4.9) by the "extended" Fibonacci number $F_i(i = 0, \pm 1, \pm 2, \pm 3, \ldots)$. Because the sum (4.29) is identically equal to 0 independently on the natural number $N$ in the expression (4.28), we came to the new fundamental property of natural numbers, which can be formulated as the following theorem.

**Theorem 4.2 ($Z$-property of natural numbers).** *If we represent the arbitrary natural number $N$ in the $\Phi$-code (4.9) and then substitute the "extended" Fibonacci numbers $F_i(i = 0, \pm 1, \pm 2, \pm 3, \ldots)$ instead of the golden ratio powers $\Phi^i(i = 0, \pm 1, \pm 2, \pm 3, \ldots)$ in the sum (4.9), then the sum that appears as a result of such substitution is identically equal to 0, independently on the initial natural number $N$, that is,*

$$\text{For any } N = \sum_i a_i \Phi^i \text{ after substitution } F_i \to \Phi^i,$$

$$\text{we have } \sum_i a_i F_i \equiv 0 (i = 0, \pm 1, \pm 2, \pm 3, \ldots).$$

$$(4.31)$$

Let's now compare the sums (4.30) and (4.9). Since the bits $a_i(i = 0, \pm 1, \pm 2, \pm 3, \ldots)$ in these sums coincide, then the sum (4.30) can be obtained from the sum (4.9) if we replace every power of the golden ratio $\Phi^i(i = 0, \pm 1, \pm 2, \pm 3, \ldots)$ in the $\Phi$-code (4.9) by the "extended" Lucas number $L_i(i = 0, \pm 1, \pm 2, \pm 3, \ldots)$. Because the sum (4.30) is identically equal to $2N$, independently on the initial natural number $N$ in the sum (4.28), we come to a new fundamental property of natural numbers, which can be formulated as the following theorem.

**Theorem 4.3 ($D$-property of natural numbers).** *If we represent the arbitrary natural number $N$ in the $\Phi$-code (4.9) and then substitute the "extended" Lucas numbers $L_i(i = 0, \pm 1, \pm 2, \pm 3, \ldots)$ instead of the golden ratio powers $\Phi^i(i = 0, \pm 1, \pm 2, \pm 3, \ldots)$ in the* *sum (4.9), then the sum that appears as a result of such substitution is identically equal to 2N, independently of the initial natural number N, that is,*

*For any* $N = \sum\limits_{i} a_i \Phi^i$ *after substitution* $L_i \to \Phi^i$,

*we have* $\sum\limits_{i} a_i L_i \equiv 2N (i = 0, \pm 1, \pm 2, \pm 3, \ldots)$.

$$(4.32)$$

## 4.4. *F*- and *L*-codes

The above discovered $Z$- and $D$-properties, given by Theorems 4.2 and 4.3, allow the introduction of the new and unusual codes for the representation of natural numbers. Because according to the $Z$-property, the sum (4.29) for the arbitrary natural numbers is identically equal to 0, that is, $B = 0$, we can write the expression (4.28) as follows:

$$N = \frac{1}{2}(A + B), \qquad (4.33)$$

where $A$ and $B$ are defined by the expressions (4.26) and (4.27), respectively.

By using the expressions (4.26) and (4.27), we can rewrite the expression (4.33) as follows:

$$N = \frac{1}{2}\left(\sum_{i} a_i L_i + \sum_{i} a_i F_i\right) = \sum_{i} a_i \frac{L_i + F_i}{2}. \qquad (4.34)$$

Taking into consideration the following well-known identity [8, 9, 11],

$$F_{i+1} = \frac{L_i + F_i}{2}, \qquad (4.35)$$

we get from (4.34) the following representation of the same natural number $N$:

$$N = \sum_{i} a_i F_{i+1}. \qquad (4.36)$$

The sum (4.36) is called *F-code of natural number N* [72].

Since the bits $a_i(i = 0, \pm1, \pm2, \pm3, \ldots)$ in the sums (4.9) and (4.36) coincide, then it follows from this fact that the *F*-code (4.36) of the arbitrary natural number $N$ can be obtained from the $\Phi$-code (4.9) of the same natural number $N$ by the replacement of the golden ratio powers $\Phi^i(i = 0, \pm1, \pm2, \pm3, \ldots)$ in the sum (4.9) by the "extended" Fibonacci number $F_{i+1}(i = 0, \pm1, \pm2, \pm3, \ldots)$, respectively, that is, $F_{i+1} \to \Phi^i$.

Let's represent now the *F*-code of $N$ (4.36) as follows:

$$N = \left( \sum_i a_i F_{i+1} \right) + 2B, \qquad (4.37)$$

where the term $B$ is defined by the sum (4.29) $B = \sum_i a_i F_i$, which is identically equal to $0 (B \equiv 0)$ according to (4.29). Then the sum (4.37) can be represented as follows:

$$N = \left( \sum_i a_i F_{i+1} \right) + 2 \left( \sum_i a_i F_i \right) = \sum_i a_i (F_{i+1} + 2F_i). \quad (4.38)$$

Taking into consideration the following well-known formula [8, 9, 11],

$$L_{i+1} = F_{i+1} + 2F_i, \qquad (4.39)$$

the sum (4.38) can be represented as follows:

$$N = \sum_i a_i L_{i+1}. \qquad (4.40)$$

The expression (4.40) is called the *L-code of the natural number N*.

Because the bits $a_i(i = 0, \pm1, \pm2, \pm3, \ldots)$ in the sums (4.9) and (4.40) coincide, it follows from this fact that the *L*-code of $N$ (4.40) can be obtained from the $\Phi$-code (4.9) of the same natural number $N$ by the replacement of the golden ratio powers

$\Phi^i(i = 0, \pm1, \pm2, \pm3, \ldots)$ in the sum (4.9) by the "extended" Lucas numbers $L_{i+1}(i = 0, \pm1, \pm2, \pm3, \ldots)$, that is, $L_{i+1} \to \Phi^i$.

It is clear that the $L$-code of $N$ (4.40) can also be obtained from the $F$-code of the same number $N$ (4.36) by the replacement of the Fibonacci numbers $F_{i+1}$ in the formula (4.36) by the Lucas numbers $L_{i+1}$, that is, $L_{i+1} \to F_{i+1}$.

Let us represent the sums (4.9), (4.36) and (4.40) in the abridged form (4.10). It is clear that the sums (4.9), (4.36) and (4.40) give three different methods of the binary representation of one and the same natural number $N$. The $\Phi$-code (4.9) is the representation of the natural number $N$ as the sum of the golden ratio powers $\Phi^i(i = 0, \pm1, \pm2, \pm3, \ldots)$, the $F$-code (4.36) is the representation of the same natural number $N$ as the sum of the "extended" Fibonacci numbers $\Phi^i(i = 0, \pm1, \pm2, \pm3, \ldots)$ and the $L$-code (4.40) is the representation of the same natural number $N$ as the sum of the "extended" Lucas numbers $L_{i+1}(i = 0, \pm1, \pm2, \pm3, \ldots)$. As mentioned above, all the sums (4.9), (4.36) and (4.40), which represent one and the same natural number $N$, have one and the same "golden" representation (4.10).

## 4.5. The Left and Right Shifts of the $\Phi$-, $F$-, and $L$-codes

Although the "golden" representations (4.10) of one and the same natural number $N$ coincide with the $\Phi$-, $F$-, and $L$-codes, the distinction among the *"golden" representations* (4.10) of these codes arises when we carry out the *left or right shifts* of the *"golden" representations* (4.10).

### 4.5.1. The shifts of the $\Phi$-code to the left and to the right

Let's denote by $N_{(k)}$ and $N_{(-k)}$ the results of the shifts of the "golden" representation (4.10) *to the left* and *to the right*, respectively. If we interpret the *"golden" representation* (4.10) as the $\Phi$-code of the natural number $N$, then its shift *to the left* (that is, to the side of the *senior* digits) on the one digit corresponds to the multiplication of the number $N$ on base $\Phi$, but its shift *to the right* (that is, to the side of the *minor* digits) on the one digit corresponds to the division of the number $N$ on base $\Phi$, that is, we obtain, respectively,

$$N_{(1)} = N \times \Phi = \sum_i a_i \Phi^{i+1}, \qquad (4.41)$$

$$N_{(-1)} = N \times \Phi^{-1} = \sum_i a_i \Phi^{i-1}. \qquad (4.42)$$

It is clear that the shift on the $k$ digits *to the left* of the *"golden" representation* (4.10) of the $\Phi$-code (4.9) corresponds to the multiplication of the number $N$ on $\Phi^k$ and the shift on the $k$ digits *to the right* corresponds to the division on $\Phi^k$ (the multiplication on $\Phi^{-k}$), that is, we obtain, respectively,

$$N_{(k)} = N \times \Phi^k = \sum_i a_i \Phi^{i+k}, \qquad (4.43)$$

$$N_{(-k)} = N \times \Phi^{-k} = \sum_i a_i \Phi^{i-k}. \qquad (4.44)$$

### 4.5.2. The shifts of the $F$-code to the left and to the right

Let's consider the shifts *to the left and to the right* of the $F$-code of the *"golden" representation* (4.10) when we interpret it as the $F$-code. If we interpret the "golden" representation (4.10) as the $F$-code, then its shift *to the left* on the $k$ digits leads us to the following expression:

$$N_{(k)} = \sum_i a_i F_{i+1+k}. \qquad (4.45)$$

Apply to the expression (4.45) the following well-known identity for the generalized Fibonacci numbers $G_k = G_{k-1} + G_{k-2}$ [11]:

$$G_{n+m} = F_{m-1} G_n + F_m G_{n+1}. \qquad (4.46)$$

For the case $G_k = F_k$, the identity (4.46) takes the following form:

$$F_{n+m} = F_{m-1}F_n + F_m F_{n+1}. \tag{4.47}$$

It follows from (4.47) that for the cases $n = i$ and $m = k + 1$, the identity (4.47) reduces to the following:

$$F_{i+1+k} = F_k F_i + F_{k+1}F_{i+1}. \tag{4.48}$$

By substituting (4.48) into the sum (4.45) and taking into consideration the $Z$-property (4.29) ($B = \sum_i a_i F_i \equiv 0$) and the value of the sum (4.36) ($N = \sum_i a_i F_{i+1}$) for the $F$-code, we obtain

$$N_{(k)} = \sum_i a_i F_{i+1+k} = \sum_i a_i (F_k F_i + F_{k+1}F_{i+1})$$

$$= F_k \sum_i a_i F_i + F_{k+1} \sum_i a_i F_{i+1}$$

$$= F_{k+1} \sum_i a_i F_{i+1} = F_{k+1} \times N. \tag{4.49}$$

If we interpret *the "golden" representation* (4.10) as the $F$-code, then the shift of (4.10) *to the right* on the $k$ digits leads us to the following sum:

$$N_{(-k)} = \sum_i a_i F_{i+1-k}. \tag{4.50}$$

Let's now consider the identity (4.47) ($F_{n+m} = F_{m-1}F_n + F_m F_{n+1}$). If we take $n = i$ and $m = -k + 1$, then we can rewrite the identity (4.47) as follows:

$$F_{i+1-k} = F_{-k}F_i + F_{-k+1}F_{i+1}. \tag{4.51}$$

By substituting (4.51) into the sum (4.50), after the simple transformations with regard to (4.29) ($B = \sum_i a_i F_i \equiv 0$) and (4.36) ($N = \sum_i a_i F_{i+1}$), we obtain

$$N_{(-k)} = \sum_i a_i F_{i+1-k} = F_{-k+1} \times N. \tag{4.52}$$

Let's now formulate the results (4.49) and (4.52) as the following theorem.

**Theorem 4.4.** *The shift to the left of the "golden" representation* (4.10), *interpreted as the F-code, on the k digits, corresponds to the multiplication of the number N by the "extended" Fibonacci number* $F_{k+1}$, *but its shift to the right on the k digits corresponds to the multiplication of the number N by the "extended" Fibonacci number* $F_{-k+1}$.

Let's now consider the formula (4.52) $(N_{(-k)} = \sum_i a_i F_{i+1-k} = F_{-k+1} \times N)$. For the case $k = 1$ (the shift to the right on the one digit), the formula (4.52) takes the following form:

$$\sum_i a_i F_i = F_0 \times N. \tag{4.53}$$

But the Fibonacci number $F_0 = 0$ and therefore the formula (4.53) reduces to the formula (4.29) $(B = \sum_i a_i F_i \equiv 0)$. This consideration is another proof of the $Z$-property given by Theorem 4.2.

Note that the shift *to the right* of the *"golden" representation* (4.10), interpreted as the $F$-code, on the three digits corresponds to the multiplication of the number $N$ on the Fibonacci number $F_{-2} = -1$. This means that such shift to the right leads us to the number $(-N) = (-1) \times N$. This property of the $F$-code together with the $Z$-property (4.31) $(B = \sum_i a_i F_i \equiv 0)$ has a number of useful applications in computer science and digital metrology.

### 4.5.3. The shifts to the left and to the right of the L-code

Let's now consider the shifts *to the left and to the right* of the *"golden" representation* (4.10), interpreted as the $L$-code (4.40). Its shifts *to the left and to the right* on the $k$ digits lead, respectively, to the following sums

$$N_{(k)} = \sum_i a_i L_{i+1+k}, \tag{4.54}$$

$$N_{(-k)} = \sum_i a_i L_{i+1-k}. \tag{4.55}$$

By using the identity (4.46) ($G_{n+m} = F_{m-1}G_n + F_m G_{n+1}$) and taking $G_k = L_k$, we can express the "extended" Lucas numbers $L_{i+1+k}$ and $L_{i+1-k}$ as follows:

$$L_{i+1+k} = F_i L_k + F_{i+1}L_{k+1}, \qquad (4.56)$$

$$L_{i+1-k} = F_i L_{-k} + F_{i+1}L_{-k+1}. \qquad (4.57)$$

Then the expressions (4.56) and (4.57) can be represented respectively, as follows:

$$N_{(k)} = \sum_i a_i L_{i+1+k} = L_k \sum_i a_i F_i + L_{k+1} \sum_i a_i F_{i+1}, \qquad (4.58)$$

$$N_{(-k)} = \sum_i a_i L_{i+1-k} = L_{-k} \sum_i a_i F_i + L_{-k+1} \sum_i a_i F_{i+1}. \qquad (4.59)$$

With regard to the expressions (4.29) ($B = \sum_i a_i F_i \equiv 0$) and (4.36) ($N = \sum_i a_i F_{i+1}$), we obtain from (4.58) and (4.59) the following results:

$$N_{(k)} = \sum_i a_i L_{i+1+k} = L_{k+1} \times N, \qquad (4.60)$$

$$N_{(-k)} = \sum_i a_i L_{i+1-k} = L_{-k+1} \times N. \qquad (4.61)$$

We can formulate the results (4.60) and (4.61) as the following theorem.

**Theorem 4.5.** *The shift to the left of the "golden" representation (4.10), interpreted as the L-code, corresponds to the multiplication of the number N by the "extended" Lucas number $L_{k+1}$, but its shift to the right on the k digits corresponds to the multiplication of the number N by the "extended" Lucas number $L_{-k+1}$.*

Let's consider the formula (4.61). For the case $k = 1$ (the shift *to the right* on the one digit), the formula (4.61) takes the following form:

$$\sum_i a_i L_i = L_0 \times N. \tag{4.62}$$

Since the "extended" Lucas number $L_0 = 2$, then the formula (4.62) reduces to the formula (4.30) ($A = \sum_i a_i L_i \equiv 2N$). This consideration is another proof of the $D$-property given by Theorem 4.3.

### 4.5.4. Numerical examples

Let's again consider the "golden" representation (4.10). We can see that the "golden" representation (4.10) is separated by the point into two parts: the left-hand part, which consists of the digits with the non-negative indices, and the right-hand part, which consists of the digits with the negative indices. As an example, we can consider the "golden" representation of the decimal number 10 in Bergman's system:

$$10_{10} = 10100.0101. \tag{4.63}$$

For the $\Phi$-code (4.9), the "golden" representation (4.63) has the following algebraic interpretation:

$$10_{10} = \Phi^4 + \Phi^2 + \Phi^{-2} + \Phi^{-4}. \tag{4.64}$$

By using the formula (4.13) ($\Phi^i = \frac{L_i + F_i \sqrt{5}}{2}$), we can represent the sum (4.64) as follows:

$$10_{10} = \Phi^4 + \Phi^2 + \Phi^{-2} + \Phi^{-4} = \frac{L_4 + F_4\sqrt{5}}{2} + \frac{L_2 + F_2\sqrt{5}}{2}$$
$$+ \frac{L_{-2} + F_{-2}\sqrt{5}}{2} + \frac{L_{-4} + F_{-4}\sqrt{5}}{2}. \tag{4.65}$$

If we take into consideration the following correlations for the "extended" Fibonacci and Lucas numbers:

$$L_{-2} = L_2; \quad L_{-4} = L_4; \quad F_{-2} = -F_2; \quad F_{-4} = -F_4,$$

then we obtain from the expression (4.65) the following result:

$$10 = \frac{2(L_4 + L_2)}{2} = L_4 + L_2 = 7 + 3.$$

Let's now consider the interpretation of the "golden" representation (4.10) as the $F$- and $L$-codes:

$$10_{10} = F_5 + F_3 + F_{-1} + F_{-3} = 5 + 2 + 1 + 2,$$
$$10_{10} = L_5 + L_3 + L_{-1} + L_{-3} = 11 + 4 - 1 - 4.$$

Also, we can check the sum (4.64) by the $Z$- and $D$-properties. If we change all the powers $\Phi^i$ in the formula (4.64) ($10_{10} = \Phi^4 + \Phi^2 + \Phi^{-2} + \Phi^{-4}$) by the "extended" Fibonacci ($F_i$) and Lucas ($L_i$) numbers, we get the following sums:

$$F_4 + F_2 + F_{-2} + F_{-4} = 3 + 1 + (-1) + (-3) = 0 \ (Z\text{-}property),$$
$$L_4 + L_2 + L_{-2} + L_{-4} = 7 + 3 + 3 + 7 = 20 = 2 \times 10 \ (D\text{-}property).$$

### 4.5.5. Algebraic summation of integers

We can use the $Z$-property to check arithmetical operations. For example, consider the operation of the algebraic summation of the two integers, $N_1 \pm N_2$. As an outcome of this operation, we always obtain a new integer. This means that the "golden" algebraic summation of the two integers, represented in the $\Phi$-, $F$- or $L$-codes, always results in the new "golden" representation of the algebraic sum $N_1 \pm N_2$ in the $\Phi$-, $F$- or $L$-codes. It follows from this consideration that *the $Z$-property is invariant about the "golden" algebraic summation.* A similar conclusion is valid for the *"golden" multiplication.* As the outcome of the "golden" division of the two integers is always the two integers, the quotient $Q$ and the remainder $R$, it follows from this consideration that *the results of the "golden" division, the integers $Q$ and $R$, preserve the $Z$-property.*

Hence, we have obtained some new fundamental properties of natural numbers, represented in the $\Phi$-, $F$- and $L$-codes. These properties (for example, the $Z$-property) are the invariants about arithmetical operations and may be used for checking arithmetical operations in computers.

## 4.6. Evaluation of Bergman's System

In conclusion, we note that Theorems 4.1–4.5 are true only for the natural numbers. This means that Theorems 4.1–4.5 give new fundamental properties of natural numbers. It is surprising for many mathematicians to know that the new mathematical properties of natural numbers, given by Theorems 4.1–4.5, were only discovered at the beginning of the 21st century, that is, 2.5 millennia after the beginning of the theoretical study of natural numbers. The golden ratio and the "extended" Fibonacci and Lucas numbers play a fundamental role in this discovery. This discovery connects together the two outstanding mathematical concepts of Greek mathematics: the *natural numbers* and the *golden ratio*. This discovery is the confirmation of the fundamental role of *Bergman's system* (4.1) for the development of the *"golden" number theory* described in [53].

### 4.6.1. The fundamental significance of Bergman's system for the development of modern mathematics and computer science

In response to the question about the importance of *Bergman's system* for the development of modern science, we can make the following bold statement: *the Bergman system is the most important modern discovery in the field of numeral systems, comparable only to the discovery of the positional principle of number representation as well as decimal and binary numeral systems.* It alters our ideas about the numeral systems, moreover, our ideas about the relation between rational and irrational numbers. In this system, the irrational number of the "golden proportion" comes to the fore; this fundamental mathematical constant becomes the basis of all numbers because with its help, any real number can be represented, and it is obvious that this mathematical result is of fundamental interest for all the number theories!

*Bergman's system* was a result of fundamental importance not only for the numeral systems but also for the number theory, but during the period of its origin (the 1950s), it was simply not noticed either by mathematicians or engineers. And the young American mathematician George Bergman estimated his mathematical discovery very modestly in his article [54]:

> *"I do not know of any useful application for numeral systems like this, except for the mental exercise and pleasant pastime, although this numeral system may be useful for the algebraic number theory."*

However, the development of computers and measuring technology refuted Bergman's pessimistic opinion regarding the practical application of Bergman's numeral system. Unlike the classical binary system, Bergman's system has *"natural"* *redundancy*, which can be effectively used to detect errors in computers.

During the 1970s and 1980s in the former Soviet Union, under the leadership of Alexey Stakhov, scientific and engineering developments, based on the Bergman numeral system [16, 17, 19, 55, 56, 62, 63, 65, 66, 71] were carried out. These developments, on the one hand, showed the exceptional efficiency of Bergman's system for designing self-correcting analog-to-digital and digital-to-analog converters and noise-resistant processors and, on the other hand, led to a new geometric definition of the *real numbers* [72, 84].

### 4.6.2. Evaluation of the applications of Bergman's system

As is known, new scientific ideas do not always arise where they are expected. Apparently, *Bergman's system* [54] is one of the most unprecedented scientific discoveries in the history of science and mathematics. First of all, it is an impressive fact that this discovery originates from the Babylonian positional numeral system with base 60 that arose at the time of the origin of mathematics. That is, Bergman's mathematical discovery [54] takes mathematics back to the initial period of its development, when the numeral systems and rules of the arithmetic operations were one of the most important problems of ancient mathematics (Babylon and ancient Egypt) [133].

This is really an unprecedented event in the history of science and mathematics. The mathematical formula (4.1) ($A = \sum_i a_i \Phi^i$) for *Bergman's system* looks so simple that it is difficult to believe that *Bergman's system* is one of the greatest modern mathematical discoveries, which has fundamental interest for the *history of mathematics*, the *number theory* and the *computer science*. In this regard, one can compare *Bergman's system* with the discovery of *incommensurable segments* made in Pythagoras' scientific school. The proof of the *incommensurability* of the diagonal and the side of the square is so simple that any amateur of mathematics can get this proof without any difficulties. However, this mathematical discovery until now causes delight because this discovery was a turning point in the development of mathematics and led to the introduction of the irrational numbers, without which it is difficult to imagine the existence of mathematics. Time will show how fair the above comparison of Bergman's system with the incommensurable segments is.

# Chapter 5

# The "Golden" Ternary Mirror-Symmetrical Arithmetic

## 5.1. The Ternary Mirror-Symmetrical Representation

### 5.1.1. Brief history

However, the ternary "golden" mirror-symmetrical arithmetic, described by Alexey Stakhov in Refs. [55, 72, 84], is the most interesting application of Bergman's system in computer science.

For the first time, the "golden" ternary mirror-symmetrical arithmetic had been described by Alexey Stakhov in the article, "Brousentsov's ternary principle, Bergman's number system and ternary mirror-symmetrical arithmetic", published in *The Computer Journal* in 2002 [72].

In order to explain the main idea of the *ternary mirror-symmetrical representation* of integers, we start from the $\Phi$-code of natural number (8.9) ($N = \sum_i a_i \Phi^i$), in which we use only the MINIMAL FORMS for the representation of integers. We recall the following fundamental property of the MINIMAL FORM:

$$
\begin{array}{|c|c|c|}
\hline
a_{k+1} & a_k & a_{k-1} \\
\hline
\mathbf{0} & \mathbf{1} & \mathbf{0} \\
\hline
\end{array}
\tag{5.1}
$$

This means that in the MINIMAL FORM, each bit $a_k = 1$ is "surrounded" by the two adjacent bits $a_{k+1} = a_{k-1} = 0$ (see in bold).

Let's now consider the following fundamental identity (4.11) for the powers of the golden ratio:

$$\Phi^{k+1} = \Phi^k + \Phi^{k-1} \ (k = 0, \pm 1, \pm 2, \pm 3, \ldots), \tag{5.2}$$

which had been called in Chapter 4 as the *additive property* of the $\Phi$-code (for the case of integers).

The identity (5.2) can be rewritten as follows:

$$\Phi^k = \Phi^{k+1} - \Phi^{k-1} \quad (k = 0, \pm 1, \pm 2, \pm 3, \ldots). \tag{5.3}$$

By using (5.3), we can carry out the following transformation over (5.1):

| | $a_{k+1}$ | $a_k$ | $a_{k-1}$ | $\Rightarrow$ | $a_{k+1}$ | $a_k$ | $a_{k-1}$ |
|---|---|---|---|---|---|---|---|
| $\Phi^k =$ | 0 | 1 | 0 | $\Rightarrow$ | 1 | 0 | $\bar{1}$ |

$$\tag{5.4}$$

where $\bar{1}$ is the negative ternary numeral, that is, $\bar{1} = -1$. It follows from (5.4) that the positive unit numeral of the $k$th digit $a_k = 1$ can be transformed into two ternary numerals: the *positive* numeral 1 of the $(k + 1)$th digit $a_{k+1} = 1$ and the *negative* numeral $\bar{1}$ of the $(k - 1)$th digit $a_{k-1} = \bar{1}$.

### 5.1.2. The conversion of the binary MINIMAL FORM into the *ternary "golden" representation* of the same integer $N$

The transformation (5.4) can be used for the conversion of the MINIMAL FORM of the *binary "golden" representation* of the natural number $N$ into the *ternary "golden" representation* of the same integer $N$.

As an example, let's now consider the *"golden" binary representation* of the natural number $N = 5$ represented in the MINIMAL FORM:

| $k$ | 4 | 3 | 2 | 1 | 0 | $-1$ | $-2$ | $-3$ | $-4$ |
|---|---|---|---|---|---|---|---|---|---|
| $N = 5$ | 0 | 1 | 0 | 0 | 0. | 1 | 0 | 0 | 1 |

$$\tag{5.5}$$

Let's convert the *"golden" binary representation* (5.5) of the natural number $N = 5$, represented in the MINIMAL FORM, into the *"golden" ternary representation* of the same natural number $N = 5$. With this purpose, we can apply the code transformation (5.4) simultaneously to all digits, being the *binary numeral* of 1 and having the *odd* indices ($k = 2m + 1$). We can see that the transformation (5.4) can be applied for the case (5.5) only for the third and ($-1$)-th digits, which are the binary numerals of 1 (see in bold). As a result of such transformations, we get the following *"golden" ternary representation* of the natural number $N = 5$:

| $k$ | 4 | 3 | 2 | 1 | 0 | $-1$ | $-2$ | $-3$ | $-4$ |
|---|---|---|---|---|---|---|---|---|---|
| $N = 5$ | 1 | 0 | $\bar{1}$ | 0 | 1. | 0 | $\bar{1}$ | 0 | 1 |

$$(5.6)$$

We can see from (5.6) that all the digits with the *odd* indices $k = 2m + 1\,(3, 1, -1, -3)$ are identically equal to 0, but the digits with the *even* indices $k = 2m\,(4, 2, -2, -4)$ take the ternary values from the set $\{\bar{1}, 0, 1\}$. This means that all the digits with the *odd* indices $k = 2m + 1\,(3, 1, -1, -3)$ are *non-informative* because their values are equal to 0 identically and they do not influence the value of the number $N = 5$. By omitting in (5.6) all the *non-informative digits*, we get the following "golden" ternary representation of the initial natural number $N = 5$:

| $k$ | 4 | 2 | 0 | $-2$ | $-4$ |
|---|---|---|---|---|---|
| $N = 5$ | 1 | $\bar{1}$ | 1. | $\bar{1}$ | 1 |

$$(5.7)$$

The algebraic interpretation of the "golden" ternary representation (5.7) has the form of the following sum:

$$N = 5 = 1 \times \Phi^4 + \bar{1} \times \Phi^2 + 1 \times \Phi^0 + \bar{1} \times \Phi^{-2} + 1 \times \Phi^{-4}$$
$$= \Phi^4 - \Phi^2 + \Phi^0 - \Phi^{-2} + \Phi^{-4}$$
$$= 6.854 - 2.618 + 1 - 0.382 + 0.146. \qquad (5.8)$$

In general, if we convert the "golden" binary representation (4.10), represented in the MINIMAL FORM, into the "golden" ternary representation, then after omitting the *non-informative digits*, we get the following sum:

$$N = \sum_i a_{2i} \Phi^{2i}, \tag{5.9}$$

where $a_{2i}$ is the ternary numeral of the $(2i)$th digit.

We can perform the following digit enumeration for the ternary numeral system (5.9). Each ternary numeral $a_{2i}$ in (5.9) is replaced by the ternary numeral $c_i$: $a_{2i} \to c_i$ $(i = 0, \pm 1, \pm 2, \pm 3, \ldots)$. As a result of such enumeration, we get the expression (5.9) in the following form:

$$N = \sum_i c_i \Phi^{2i}, \tag{5.10}$$

where $c_i \in \{\bar{1}, 0, 1\}$ is the ternary numeral of the $i$th digit, $\Phi^{2i}$ is the weight of the $i$th digit, $\Phi^2 = 2.618$ is the *base* of the numeral system (5.10). We name the sum (5.10) the *ternary $\Phi$-code* of natural number $N$.

With regard to the expression (5.10), the *"golden" ternary representation* (5.7) of the natural number $N = 5$ takes the following form:

| $k$ | 2 | 1 | 0 | −1 | −2 |
|-----|---|---|----|----|----|
| $N = 5$ | 1 | $\bar{1}$ | 1. | $\bar{1}$ | 1 |

$$(5.11)$$

The conversion of the *binary $\Phi$-code* (5.5) of the *natural number N* into the *"golden" ternary representation* (5.10) of the same *natural number N* can be carried out by means of a simple combinative logic circuit, which transforms the 3-bit binary MINIMAL FORM $(a_{k+1} a_k a_{k-1})$ into the *ternary informative digit* $c_i$ of the ternary "golden" representation in accordance with Table 5.1.

Table 5.1. Conversion of the binary MINIMAL FORM $a_{2i+1}a_{2i}a_{2i-1}$ into the ternary numeral $c_{2i}$.

| $a_{2i+1}$ | $a_{2i}$ | $a_{2i-1}$ | $\Rightarrow$ | $c_{2i}$ | | $a_{2i+1}$ | $a_{2i}$ | $a_{2i-1}$ | $\Rightarrow$ | $c_{2i}$ |
|---|---|---|---|---|---|---|---|---|---|---|
| 0 | 0 | 0 | $\Rightarrow$ | 0 | | $0 \to 0$ | 0 | $0 \leftarrow 0$ | $\Rightarrow$ | $0+0+0 = 0$ |
| 0 | 0 | 1 | $\Rightarrow$ | 1 | | $0 \to 0$ | 0 | $1 \leftarrow 1$ | $\Rightarrow$ | $0+0+1 = 1$ |
| 0 | 1 | 0 | $\Rightarrow$ | 1 | $\Rightarrow$ | $0 \to 0$ | 1 | $0 \leftarrow 0$ | $\Rightarrow$ | $0+1+0 = 1$ |
| 1 | 0 | 0 | $\Rightarrow$ | $\bar{1}$ | | $1 \to \bar{1}$ | 0 | $0 \leftarrow 0$ | $\Rightarrow$ | $\bar{1}+0+0 = \bar{1}$ |
| 1 | 0 | 1 | $\Rightarrow$ | 0 | | $1 \to \bar{1}$ | 0 | $1 \leftarrow 1$ | $\Rightarrow$ | $\bar{1}+0+1 = 0$ |

First of all, we note that Table 5.1 uses only the five neighboring binary code combinations $a_{2i+1}a_{2i}a_{2i-1}$ from eight of the possible 3-bit binary code combinations because the initial *"golden" binary representation* of the kind (5.5) is represented in the MINIMAL FORM and therefore the neighboring binary code combinations $a_{2i+1}a_{2i}a_{2i-1} \in \{011, 110, 111\}$ are *prohibited* for the MINIMAL FORM.

The feature of transformation of the code combination, consisting of the three adjacent binary numerals $a_{2i+1}a_{2i}a_{2i-1}$, presented in the MINIMAL FORM, into the *ternary numeral* $c_{2i} \in \{\bar{1}, 0, 1\}$, that is, $a_{2i+1}a_{2i}a_{2i-1} \Rightarrow c_{2i}$, consists of the following. The *binary numeral* $a_{2i}$ ($i = 0, \pm1, \pm2, \pm3, \ldots$) of the binary 3-bit code combination $a_{2i+1}a_{2i}a_{2i-1}$ is its "key" numeral, where we should place the *"treat"* $c_{2i} \in \{\bar{1}, 0, 1\}$, which means the *unit* of the *information measure* in the ternary code.

The code transformations (000 $\to$ 0 and 010 $\to$ 1), taken from the second and fourth rows of Table 5.1, respectively, are trivial. The code transformations (001 $\to$ 1, 100 $\to$ $\bar{1}$, $\bar{1}$01 $\to$ 0), taken from the third, fifth and sixth rows of Table 5.1, require more detailed explanation.

Really, the code transformation (001 $\to$ 1), taken from the third row of Table 5.1, means that the positive ternary numeral (1), arising in accordance with (5.4) from the *right-hand binary digit* $a_{2i+1} = 1$, is summarized with the *"zero" numeral* (0) from the *left-hand binary digit* $a_{2i-1} = 0$, that is, $1 + 0 + 0 = 1$.

The code transformation of the sixth row $\bar{1}$01 $\to$ 0 means that the *negative numeral* ($\bar{1}$), arising in accordance with (5.4) from the *left-hand binary digit* $a_{2i+1} = 1$, is summarized with the *positive numeral* (1), arising from the *right-hand binary digit* $a_{2i-1} = 1$. It follows from this consideration that their sum ($\bar{1} + 1$) equals to the *ternary numeral* $c_{2i} = 0$.

## 5.2. The "Golden" Ternary $F$- and $L$-representations

In the previous section, we introduced the so-called $F$-and $L$-codes (4.36) and (4.40) for the *Bergman system* (4.9). Recall that these unusual codes are the equivalents of the $\Phi$-code (4.9) of the same *natural number* $N$. By using the ternary $\Phi$-code of the *natural number* $N$, given by (5.10), it is easy to write the ternary $F$-and $L$-codes of the same *natural number* $N$ in the following forms:

$$N = \sum_i c_i F_{2i+1}, \tag{5.12}$$

$$N = \sum_i c_i L_{2i+1}. \tag{5.13}$$

Note that the values of the ternary digits in the codes (5.10), (5.12), (5.13) coincide. It follows from this consideration that the "golden" ternary $\Phi$-, $F$-, $L$-representations of the number $N = 5$, given by the example (5.11), have three different algebraic interpretations:

(a) *The ternary* $\Phi$-*code*

$$5 = 1 \times \Phi^4 + \bar{1} \times \Phi^2 + 1 \times \Phi^0 + \bar{1} \times \Phi^{-2} + 1 \times \Phi^{-4}$$

$$= \Phi^4 - \Phi^2 + \Phi^0 - \Phi^{-2} + \Phi^{-4}$$

$$= \frac{L_4 + F_4\sqrt{5}}{2} - \frac{L_2 + F_2\sqrt{5}}{2} + 1 - \frac{L_{-2} + F_{-2}\sqrt{5}}{2}$$

$$+ \frac{L_{-4} + F_{-4}\sqrt{5}}{2} = L_4 - L_2 + 1 = 7 - 3 + 1.$$

(b) *The ternary* $F$-*code*

$$5 = 1 \times F_5 + \bar{1} \times F_3 + 1 \times F_1 + \bar{1} \times F_{-1} + 1 \times F_{-3} = 5 - 2 + 1 - 1 + 2.$$

(c) *The ternary* $L$-*code*

$$5 = 1 \times L_5 + \bar{1} \times L_3 + 1 \times L_1 + \bar{1} \times L_{-1} + 1 \times L_{-3} = 11 - 4 + 1 + 1 - 4.$$

Note that we have used for the calculation of the above "golden" ternary $\Phi$-, $F$- and $L$-representations the properties (4.8) $F_{-n} = (-1)^{n+1}F_n$; $L_{-n} = (-1)^n L_n$ and Table 4.1 of Chapter 4.

### 5.2.1. The representation of negative numbers

Similar to the ternary-symmetrical numeral system, $N = \sum_{i=0}^{n-1} b_i 3^i$, the important advantage of the ternary numeral system (5.10) is a possibility to represent both positive and negative numbers in a "direct" code. The "golden" ternary representation of the negative number $(-N)$ can be obtained from the "golden" ternary $\Phi$-representation of the initial natural number $N$ by using the "*ternary inversion*" *rule* $(\bar{1} \to 1; 0 \to 0; 1 \to \bar{1})$. By using this rule to the "golden" ternary $\Phi$-representation (5.11) of number 5, we get the following "golden" ternary $\Phi$-representation of the negative number $(-5)$:

| $k$ | 2 | 1 | 0 | $-1$ | $-2$ |
|---|---|---|---|---|---|
| $\downarrow N = 5$ | 1 | $\bar{1}$ | 0 | $\bar{1}$ | 1 |
| $N = -5$ | $\bar{1}$ | 1 | 0 | 1 | $\bar{1}$ |

$$(5.14)$$

## 5.3. Mirror-Symmetrical Property: The Base and Comparison of the "Golden" Ternary Representations and Their Redundancy

### 5.3.1. Mirror-symmetrical property

By considering the "golden" ternary $\Phi$-representation (5.11) of the number $N = 5 = 1\bar{1}1.\bar{1}1 = \Phi^4 - \Phi^2 + \Phi^0 - \Phi^{-2} + \Phi^{-4}$, we can see that the left-hand part $(1\bar{1})$ of the "golden" ternary $\Phi$-representation (5.11) is *mirror-symmetrical* to its right-hand part $(\bar{1}1)$ relatively to the zeroth digit with the weight $\Phi^0 = 1$. This property, named the *mirror-symmetrical property* of the numeral system (5.11) [72], is a fundamental property of the "*golden*" *ternary* $\Phi$-, *F- and L-representations* of integers (positive and negative). Table 5.2 demonstrates this property for some initial natural numbers.

Let's give the explanations of Table 5.2. The first row $i$ means the *digit indices* of the seven-digit "*golden*" *ternary mirror-symmetrical code* (5.10), the second row $\Phi^{2i}$ means the digit weights of the seven-digit "*golden*" *ternary mirror-symmetrical* $\Phi$-*code* (5.10), the third row $F_{2i+1}$ means the digit weights of the seven-digit "*golden*" *ternary mirror-symmetrical F-code* (5.12), the fourth row $L_{2i+1}$ means the digit weights of the seven-digit "*golden*" *ternary mirror-symmetrical L-code* (5.13). The rows from 6 to 16 $N$ show the "*golden*" *ternary mirror-symmetrical representations* of the positive integers from

Table 5.2. Property of the "mirror symmetry".

| $i$ | 3 | 2 | 1 | **0** | $-1$ | $-2$ | $-3$ |
|---|---|---|---|---|---|---|---|
| $\Phi^{2i}$ | $\Phi^6$ | $\Phi^4$ | $\Phi^2$ | $\Phi^0$ | $\Phi^{-2}$ | $\Phi^{-4}$ | $\Phi^{-6}$ |
| $F_{2i+1}$ | 13 | 5 | 2 | 1 | 1 | 2 | 5 |
| $L_{2i+1}$ | 29 | 11 | 4 | 1 | $-1$ | $-4$ | $-11$ |
| $N\downarrow$ | $\downarrow$ | $\downarrow$ | $\downarrow$ | $\downarrow$ | $\downarrow$ | $\downarrow$ | $\downarrow$ |
| 0 | 0 | 0 | 0 | 0. | 0 | 0 | 0 |
| 1 | 0 | 0 | 0 | 1. | 0 | 0 | 0 |
| 2 | 0 | 0 | 1 | $\overline{1}$. | 1 | 0 | 0 |
| 3 | 0 | 0 | 1 | 0. | 1 | 0 | 0 |
| 4 | 0 | 0 | 1 | 1. | 1 | 0 | 0 |
| 5 | 0 | 1 | $\overline{1}$ | 1. | $\overline{1}$ | 1 | 0 |
| 6 | 0 | 1 | 0 | $\overline{1}$. | 0 | 1 | 0 |
| 7 | 0 | 1 | 0 | 0. | 0 | 1 | 0 |
| 8 | 0 | 1 | 0 | 1. | 0 | 1 | 0 |
| 9 | 0 | 1 | 1 | $\overline{1}$. | 1 | 1 | 0 |
| 10 | 0 | 1 | 1 | 0. | 1 | 1 | 0 |

0 to 10. The analysis of the "golden" ternary representations of the integers 0–10 in Table 5.3 shows that all the "golden" ternary representations of the positive integers from 0 to 10 have a unique mathematical property called the *mirror-symmetrical property of integers*. Based on this fundamental property, the "ternary numeral system", given by (5.10), was named the *ternary mirror-symmetrical numeral system* [72].

Another interesting feature of the *"golden" ternary mirror-symmetrical system* (5.10) follows from Table 5.2. For all the canonical numeral systems $x = \sum_{i=0}^{n-1} b_i R^i$, the "extension" of the positional representation of the number is carried out only on the side of the senior digits. For the *"golden" ternary mirror-symmetrical system* (5.10), the "extension" of the *ternary mirror-symmetrical representation* occur on both sides, i.e., on the sides of the senior and minor digits simultaneously. This feature and also the property of the *mirror-symmetry* and other features single out the ternary mirror-symmetrical positional numeral system (5.10) among all other positional numeral systems.

## 5.3.2. The base of the "golden" ternary mirror-symmetrical numeral system

It follows from (5.10) that the base of the *"golden" ternary numeral system* (5.10) is the *square of the golden ratio*, that is,

$$\Phi^2 = \frac{3 + \sqrt{5}}{2} \approx 2.618. \tag{5.15}$$

This means that the *"golden" ternary numeral system* (5.10) is the *numeral system with an irrational base*.

The base (5.15) has the following traditional representation:

$$\Phi^2 = 10 = 1 \times \Phi^2 + 0 \times \Phi^0.$$

## 5.3.3. Comparison of the "golden" ternary mirror-symmetrical numbers

Let's consider the set of weights of the $(2n + 1)$th digit "golden" ternary mirror-symmetrical $\Phi$-code (5.10):

$$\{\Phi^{2n}, \Phi^{2(n-1)}, \ldots, \Phi^4, \Phi^2, \Phi^0, \Phi^{-2}, \Phi^{-4}, \ldots, \Phi^{-(2n-1)}, \Phi^{-2n}\}.$$

$$\tag{5.16}$$

It is easy to prove that the weight of the $n$th digit of the *"golden" ternary mirror-symmetrical system* (5.10) is always strictly more than the sum of the remaining weights of (5.10) at the right of the $n$th digit. It follows from this fact that the higher significant digit of the "golden" ternary mirror-symmetrical $\Phi$-code (5.10) contains in itself the information about the sign of the ternary mirror-symmetrical number. If the numeral of the senior significant digit of the *"golden" ternary mirror-symmetrical $\Phi$-code* (5.10) is equal to 1, this means that the *"golden" ternary mirror-symmetrical number* is positive regardless of the weights of the remaining digits of the *"golden" mirror-symmetric numeral system*. If the numeral of the *senior significant digit* of the *"golden" ternary mirror-symmetrical $\Phi$-code* (5.10) is equal to $\bar{1}$, this means that the *"golden" ternary mirror-symmetrical number* is negative.

A very simple method for the comparison of the two "golden" ternary mirror-symmetrical numbers $A$ and $B$ follows from this consideration. The comparison begins from the senior digits of the comparable "golden" mirror-symmetrical numbers and continues until we obtain the first pair of the non-coincident ternary mirror-symmetrical numerals $a_k$ and $b_k$. If the numeral $a_k > b_k (1 > 0, 1 > \bar{1}, 0 > \bar{1})$, then $A > B$. In the opposite case, $A < B$.

Hence, we have found the two important advantages of the *"golden" ternary mirror-symmetrical numeral system* (5.10):

(1) Similar to the classical ternary-symmetrical numeral system $N = \sum_{i=0}^{n-1} b_i 3^i$, the sign of the "golden" ternary mirror-symmetrical number is determined by the sign of the senior significant (1 or $\bar{1}$) numeral of the *"golden" ternary mirror-symmetrical numeral system* (5.10).

(2) The comparison of the "golden" ternary mirror-symmetrical numbers is carried out similar to the comparison of the ternary numbers, represented in the classical *ternary-symmetrical numeral system* $N = \sum_{i=0}^{n-1} b_i 3^i$, that is, by starting from the senior "golden" ternary digits in the direction to the minor digits until we obtain the first pair of the non-coincident "golden" ternary mirror-symmetrical digits.

### 5.3.4. The range of the ternary mirror-symmetrical representation of numbers

Let's now consider the range of number representation in the numeral system (5.10). Suppose that the "golden" ternary mirror-symmetrical $\Phi$-code (5.10) has $2m+1$ ternary digits. In this case, by using (5.10), we can represent all integers (positive and negative) in the range from

$$N_{\max} = \underbrace{1\ 1\ \ldots\ 1}_{m}\ 1.\ \underbrace{1\ 1\ \ldots\ 1}_{m} \tag{5.17}$$

to

$$N_{\min} = \underbrace{\bar{1}\ \bar{1}\ \ldots\ \bar{1}}_{m}\ \bar{1}.\ \underbrace{\bar{1}\ \bar{1}\ \ldots\ \bar{1}}_{m}. \tag{5.18}$$

It is clear that $N_{max}$ is a positive number and $N_{min}$ is a negative number.

It is clear that $N_{min}$ is the ternary inversing of $N_{max}$, that is, they are equal by absolute value:

$$|N_{min}| = N_{max}. \tag{5.19}$$

It follows from this consideration that by using the $2m + 1$ "golden" ternary mirror-symmetrical $\Phi$-code (5.10), we can represent

$$2N_{max} + 1 \tag{5.20}$$

"golden" ternary mirror-symmetrical integers (positive and negative) by including the number of 0.

For the calculation of $N_{max}$, we can interpret (5.17) as the "golden" ternary mirror-symmetrical $L$-code (5.13). Taking into consideration the expression (5.13) for the $L-$code, we can give the following algebraic interpretation of (5.17):

$$N_{max} = L_{2m+1} + L_{2m-1} + \cdots + L_5 + L_3 + L_1 + L_{-1}$$
$$+ L_{-3} + L_{-5} + \cdots + L_{-(2m-1)}. \tag{5.21}$$

It follows from Table 1.4 (see Chapter 1 of Volume I) that for the following "extended" Lucas numbers

| $n$ | 0 | 1 | 2 | 3 | 4 | 5 | 6 | 7 | 8 | 9 | 10 |
|------|---|----|---|----|---|-----|----|-----|----|-----|-----|
| $L_n$ | 2 | 1 | 3 | 4 | 7 | 11 | 18 | 29 | 47 | 76 | 123 |
| $L_{-n}$ | 2 | -1 | 3 | -4 | 7 | -11 | 18 | -29 | 47 | -76 | 123 |

we have the following simple relation between the Lucas numbers $L_{-i}$ and $L_i$:

$$L_{-i} = (-1)^i L_i \ (i = 0, \pm 1, \pm 2, \pm 3, \ldots). \tag{5.22}$$

This means that for the *odd* indices $i = 1, 3, 5, \ldots, 2k - 1$, we have the following property for the Lucas numbers [28]:

$$L_{-(2k-1)} = -L_{2k-1} \ (k = 1, 2, 3, \ldots). \tag{5.23}$$

Taking into consideration the property (5.23), we can rewrite the sum (5.21) as follows:

$$N_{\max} = L_{2m+1} + L_{2m-1} + \cdots + L_5 + L_3$$
$$+ L_1 - L_1 - L_3 - L_5 - \cdots - L_{2m-1}, \qquad (5.24)$$

from where we get the following result:

$$N_{\max} = L_{2m+1}. \qquad (5.25)$$

Taking into consideration (5.20) and (5.25), we can formulate the following theorem.

**Theorem 5.1.** *By using* $(2m + 1)$ *ternary mirror-symmetrical digits, using the ternary numerals* $\{\bar{1}, 0, 1\}$*, we can represent in the ternary mirror-symmetrical numeral system* (5.10) *the* $2L_{2m+1} + 1$ *integers (including the* $L_{2m+1}$ *positive integers, the* $L_{2m+1}$ *negative integers and the number of* 0*) in the range from* $(-L_{2m+1})$ *to* $(+L_{2m+1})$*, where* $L_{2m+1}$ *is the Lucas number.*

### 5.3.5. Code redundancy of the ternary mirror-symmetrical numeral system

The $(2m + 1)$-digit "golden" ternary mirror-symmetrical $\Phi$-code (5.10) ($N = \sum_i c_i \Phi^{2i}$) is the redundant ternary numeral system. For the calculation of the *code redundancy* of the $(2m+1)$-digit "golden" ternary mirror-symmetrical $\Phi$-code (5.10), we compare the ranges of number representations for the $(2m + 1)$-digit *"golden" mirror-symmetrical ternary* $\Phi$-*code* (5.10) and the ternary-symmetrical numeral system $N = \sum_{i=0}^{n-1} b_i 3^i$, which is the non-redundant ternary numeral system.

For the calculation of the relative redundancy $R$, we will use the following formula:

$$R = \frac{k - n}{k} = 1 - \frac{n}{k}, \qquad (5.26)$$

where $k$ and $n$ are the digit numbers of the "golden" redundant ternary mirror-symmetrical numeral system (5.10) ($N = \sum_i c_i \Phi^{2i}$) and the non-redundant ternary-symmetrical numeral systems

$N = \sum_{i=0}^{n-1} b_i 3^i$ for the representation of the same range of numbers, respectively.

According to Theorem 5.3, the $(2m + 1)$-digit "golden" ternary mirror-symmetrical $\Phi$-code (5.10) ($N = \sum_i c_i \Phi^{2i}$) can represent ($2L_{2m+1}+1 \approx 2L_{2m+1}$) integers in the range from $-L_{2m+1}$ to $L_{2m+1}$, where $L_{2m+1}$ is the Lucas number.

We show in Chapter 2 of Volume I that the non-redundant $n$-digit ternary-symmetrical code $N = \sum_{i=0}^{n-1} b_i 3^i$ can represent the $3^n$ integers in the range from $N_{\min} = -\frac{3^n-1}{2}$ up to $N_{\max} = \frac{3^n-1}{2}$. For the calculation of the code redundancy of the "golden" ternary mirror-symmetrical $\Phi$-code (5.10), we have to equate the ranges of number representation of these two ternary codes, that is.

$$3^n \approx 2L_{2m+1} \qquad (5.27)$$

and then calculate the number of digit $n$ necessary for the representation of the range $2L_{2m+1}$. To solve this task, we express the Lucas number $L_{2m+1}$ through the golden ratio $\Phi = \frac{1+\sqrt{5}}{2}$ by using Binet's formula for the Lucas numbers. By using Binet's formula

$$L_n = \begin{cases} \Phi^n + \Phi^{-n} & \text{for } n = 2k, \\ \Phi^n - \Phi^{-n} & \text{for } n = 2k + 1, \end{cases}$$

we can write the following approximate formula for the calculation of the Lucas number $L_{2m+1}$:

$$L_{2m+1} \approx \Phi^{2m+1}. \qquad (5.28)$$

By substituting (5.28) into (5.27), we get the following approximate formula for the comparison of the ranges:

$$3^n \approx 2\Phi^{2m+1}. \qquad (5.29)$$

If we take the logarithm with base 3 from both parts of the equality (5.29), we can get the following formula for the calculation of $n$:

$$n = (2m + 1) \log_3 \Phi + \log_3 2 \approx (2m + 1) \log_3 \Phi. \qquad (5.30)$$

The expression (5.30) can be used for the calculation of the code redundancy of the *"golden" ternary mirror-symmetrical* $\Phi$-*code* (5.10). For this purpose, we substitute into the formula (5.26) the following expressions instead of $k$ and $n$:

$$k = 2m + 1 \quad \text{and} \quad n = (2m + 1)\log_3 \Phi. \tag{5.31}$$

As a result, we get the following value for the relative redundancy of the *"golden" ternary mirror-symmetrical* $\Phi$-*code* (5.10):

$$R = 1 - \frac{n}{k} = 1 - \frac{(2m + 1)\log_3 \Phi}{(2m + 1)} = 1 - \log_3 \Phi \approx 0.562\,(56.2\%). \tag{5.32}$$

## 5.4. The Ternary Mirror-Symmetrical Summation and Subtraction

### 5.4.1. Mirror-symmetrical summation-subtraction

The following identities for the golden ratio powers underlie the *"golden" ternary mirror-symmetric summation-subtraction*:

$$2\Phi^{2k} = \Phi^{2(k+1)} - \Phi^{2k} + \Phi^{2(k-1)}, \tag{5.33}$$

$$3\Phi^{2k} = \Phi^{2(k+1)} + 0 + \Phi^{2(k-1)}, \tag{5.34}$$

$$4\Phi^{2k} = \Phi^{2(k+1)} + \Phi^{2k} + \Phi^{2(k-1)}, \tag{5.35}$$

where $k = 0, \pm 1, \pm 2, \pm 3, \ldots$.

The identity (5.33) underlies the *"golden" ternary mirror-symmetrical summation–subtraction* of the two *"golden" single-digit ternary mirror-symmetrical digits* and gives *the rule of the carryover and intermediate sum formation* (Table 5.3).

Table 5.3. The "golden" ternary mirror-symmetrical summation–subtraction.

| $a_k/b_k$ | $\bar{1}$ | 0 | 1 |
|---|---|---|---|
| $\bar{1}$ | $\bar{1}1\bar{1}$ | $\bar{1}$ | 0 |
| 0 | $\bar{1}$ | 0 | 1 |
| 1 | 0 | 1 | $1\bar{1}1$ |

The main peculiarity of Table 5.3 appears at the summation (subtraction) of the two ternary numerals $a_k + b_k$ with equal signs, i.e.,

$$
\begin{array}{ccccccc}
a_k & + & b_k & = & c_k & s_k & c_k \\
1 & + & 1 & = & 1 & \bar{1} & 1 \\
\bar{1} & + & \bar{1} & = & \bar{1} & 1 & \bar{1}
\end{array}
$$

where $s_k$ is the *intermediate sum* and $c_k$ is the carryovers from the $k$th digit.

We can see that at the *"golden" ternary mirror-symmetrical summation (subtraction)* of the ternary single-digit numerals $a_k + b_k$ of the same sign, there arises the intermediate sum $s_k$ of the *opposite sign* and the carryover $c_k$ of the *same sign*. However, the carryovers from the $k$th digit spreads simultaneously to the two adjacent digits, namely, to the *adjacent left-hand*, that is, the $(k + 1)$th ternary digit, and to the *adjacent right-hand*, that is, the $(k - 1)$th digit. *Symmetrical appearance of the left-hand and right-hand carryovers to the side of the adjacent left-hand and right-hand digits is the main peculiarity of the "golden" ternary mirror-symmetrical summation (subtraction).*

Table 5.4 describes the functioning of the simplest "golden" ternary mirror-symmetrical summator (subtractor) called the *"golden" single-digit ternary mirror-symmetrical half-summator (subtractor)*. This half-summator (subtractor) is a combinative logic circuit, which has two ternary inputs $a_k$ and $b_k$ and two ternary outputs $s_k$ and $c_k$ and operates according to Table 5.3 (Fig. 5.1(a)).

Note that we can use the single-digit ternary-symmetrical half-summator in Fig. 2.11 (see Chapter 2 of Volume I) as the single-digit ternary mirror-symmetrical half-summator $2\Sigma$ in Fig. 5.1(a).

Now, describe the logical functioning of the *"golden" ternary mirror-symmetrical full single-digit summator (subtractor)* of the kind $4\Sigma$. First of all, we note that the number of all the possible four-digit ternary input combinations of the "golden" mirror-symmetrical full summator (subtractor) in Fig. 5.1(b) is equal to $3^4 = 81$. The values of the output variables $s_k$ and $c_k$ are some discrete functions of the algebraic sum $S$ of the input ternary variables $a_k$, $b_k$, $c_{k-1}$, $c_{k+1}$, that is,

$$S = a_k + b_k + c_{k-1} + c_{k+1}. \tag{5.36}$$

The sum (5.36) takes the values from the set $\{\bar{4}, \bar{3}, \bar{2}, \bar{1}, 0, 1, 2, 3, 4\}$. The functioning of the mirror-symmetrical full summator (subtractor) of the kind $4\Sigma$ (Fig. 5.1(b)) is as follows; the summator (subtractor) forms the output ternary code combinations $c_k s_k$ in accordance with the value of the sum (5.36):

$$\bar{4} = \bar{1}\,\bar{1}; \; \bar{3} = \bar{1}0; \; \bar{2} = \bar{1}1; \; \bar{1} = 0\bar{1}; \; 0 = 00;$$

$$1 = 01; \; 2 = 1\bar{1}; \; 3 = 10; \; 4 = 11.$$

The output numerals $c_k s_k$ of the **"golden" ternary mirror-symmetrical single-digit summators (subtractors)** in Fig. 5.1(b) are the values of the intermediate sum $s_k$ and the senior numerals are the values of the carryover $c_k$, which spread to the adjacent (left-hand and right-hand) digits.

*Let's note that the arithmetical operations of the summation and the subtraction are the same arithmetical operations in the "golden" ternary mirror-symmetrical arithmetic.* We use the same structures of the summators and the subtractors in Fig. 5.1 for their fulfillment.

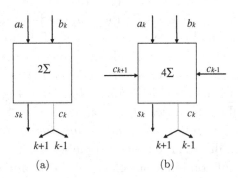

(a)          (b)

Fig. 5.1. The "golden" ternary mirror-symmetrical single-digit summators (subtractors): (a) half-summator (subtractor); (b) full summator (subtractor).

## 5.4.2. Examples of the ternary mirror-symmetrical summation (subtraction)

**Example 5.1.** Summarize the two "golden" ternary mirror-symmetrical numbers $5 + 10$:

$$
\begin{array}{rcccccccc}
5 & = & 0 & 1 & \bar{1} & 1. & \bar{1} & 1 & 0 \\
10 & = & 0 & 1 & 1 & 0. & 1 & 1 & 0 \\
s_1 & = & 0 & \bar{1} & 0 & 1. & 0 & \bar{1} & 0 \\
c_1 & & 1 & \leftrightarrow & 1 & \downarrow & 1 & \leftrightarrow & 1 \\
\hline
15 & = & 1 & \bar{1} & 1 & 1. & 1 & \bar{1} & 1
\end{array}
$$

Let's note that the symbol $\leftrightarrow$ here marks the process of the spreading of the carryover.

We can see that the summation for this example consists of two steps. The *first step* involves the formation of the first multi-digit intermediate sum $s_1$ and the first multi-digit carryover $c_1$ according to Table 5.3. The *second step* involves the summation of the numbers $s_1 + c_1$ according to Table 5.4. Since for this case, the second multi-digit intermediate carryover $c_2 = 0$, then the summation is over and the sum $s_1 + c_1 = 15$ is the summation result. It is important to emphasize that the summation result

$$15 = 1\,\bar{1}\,11.1\,\bar{1}\,1 \tag{5.37}$$

is represented in the *"golden" ternary mirror-symmetrical form.*

As we note above, the important advantage of the *"golden" ternary mirror-symmetrical numeral system* (5.10) is a possibility to summarize all integers (positive and negative) in the *"direct"* code, that is, without using the notions of the *"inverse"* and *"additional"* codes.

**Example 5.2.** Subtract the negative ternary mirror-symmetrical number $(-24)$ from the positive ternary mirror-symmetrical number 15. The ternary mirror-symmetrical subtraction is carried out as the ternary mirror-symmetrical summation of the ternary mirror-symmetrical numbers $(-24) + 15$:

$$
\begin{array}{rccccccc}
-24 = & \bar{1} & \bar{1} & 0 & 1. & 0 & \bar{1} & \bar{1} \\
15 = & 1 & \bar{1} & 1 & 1. & 1 & \bar{1} & 1 \\
s_1 & 0 & 1 & 1 & \bar{1}. & 1 & 1 & 0 \\
c_1^1 = & & \downarrow & 1 & \leftrightarrow & 1 & \downarrow & \\
c_1^2 = & \bar{1} & \leftrightarrow & \bar{1} & & \bar{1} & \leftrightarrow & \bar{1} \\
-9 = & \bar{1} & 1 & 1 & \bar{1}. & 1 & 1 & \bar{1}
\end{array}
$$

We can see that the ternary mirror-symmetrical summation of numbers $(-24) + 15$ consists of two steps for the given case. The *first step* involves the formation of the first multi-digit intermediate sum $s_1$ and the first multi-digit carryover $c_1 = c_1^1 + c_1^2$ according to Table 5.3. The *second step* involves the summarizing of the numbers $s_1 + c_1^1 + c_1^2$. Here, we use the functioning rule of the "golden" ternary mirror-symmetrical single-digit summator (subtractor) in Fig. 5.1(b). Since for this case, the second multi-digit intermediate carryover $c_2 = 0$, then the summation (subtraction) is over and the sum $s_1 + c_1^1 + c_1^2 = -9$ is the result of the summation (subtraction). It is important to emphasize that the result of summation (subtraction)

$$-9 = \bar{1}\,11\,\bar{1}.11\,\bar{1} \tag{5.38}$$

is represented in the *"golden" ternary mirror-symmetrical form*.

   The most important observation from this example consists of the fact that we used Table 5.3 to summarize the negative number $(-24)$ with the positive number 15. But the summation of the two numbers $(-24) + 15$ actually reduces to their subtraction. *This means that the concept of subtraction as specific arithmetic operation in the "golden" ternary mirror-symmetrical arithmetic can be abolished.* We simply summarize the two "golden" ternary mirror-symmetrical numbers, which can be of the same or opposite signs. For the "golden" ternary mirror-symmetrical arithmetic, the signs of the "golden" ternary mirror-symmetrical numbers do not have any significance.

## 5.5.  Ternary Mirror-Symmetrical Multi-Digit Summator (Subtractor)

The multi-digit combinative mirror-symmetrical summator (subtractor) (Fig. 5.2), which carries out the summation (subtraction) of the two $(2m + 1)$-digit *"golden" ternary mirror-symmetrical numbers*, is the combinative logic circuit, which consists of the $2m + 1$ *single-digit "golden" ternary mirror-symmetrical summators* of the kind $4\Sigma$ (Fig. 5.1(b)).

We can see from Fig. 5.2 that the main peculiarity of the *multi-digit "golden" ternary mirror-symmetrical summator (subtractor)* consists in the fact that the carryover from each digit is spreading symmetrically to the adjacent *left-hand* and the *right-hand* digits. The two "golden" ternary mirror-symmetrical numbers $A$ and $B$ enter the multi-digit input of the summator (subtractor). The single-digit summator (subtractor) $4\Sigma_0$ separates the summator (subtractor) in Fig. 5.2 into two symmetrical parts: the single-digits summators (subtractors) $4\Sigma_1$, $4\Sigma_2$, $4\Sigma_3$ for the senior digit and the single-digit summators (subtractors) $4\Sigma_{-1}$, $4\Sigma_{-2}$, $4\Sigma_{-3}$ for the minor digits.

Note that the "golden" ternary multi-digit mirror-symmetrical summator (subtractor) in Fig. 5.2 carries out the operations of the ternary mirror-symmetrical summation (subtraction) of the "golden" ternary multi-digit mirror-symmetrical numbers. This means that the "golden" ternary multi-digit mirror-symmetrical summation of the two positive mirror-symmetrical numbers with the same signs $5 + 10$ (Example 5.1) and the two "golden" ternary mirror-symmetrical numbers with the different signs $(-24) + 15$

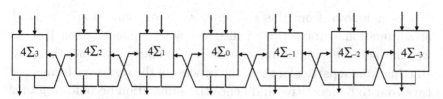

Fig. 5.2. The "golden" ternary mirror-symmetrical multi-digit summator (subtractor).

(Example 5.2) can be carried out by using the same "golden" ternary mirror-symmetrical multi-digit summator, which is also the "golden" ternary mirror-symmetrical multi-digit subtractor.

### 5.5.1. "Swing"-phenomenon

Let's now summarize the two equal numbers $5 + 5$ represented in the "golden" ternary mirror-symmetrical numeral system (5.10):

$$
\begin{array}{ccccccccc}
5 & = & 0 & 1 & \bar{1} & 1. & \bar{1} & 1 & 0 \\
5 & = & 0 & 1 & \bar{1} & 1. & \bar{1} & 1 & 0 \\
\hline
 & & 0 & \bar{1} & 1 & \bar{1}. & 1 & \bar{1} & 0 \\
 & & & \downarrow & 1 & \leftrightarrow & 1 & \downarrow & \\
 & & 1 & \leftrightarrow & 1 & & 1 & \leftrightarrow & 1 \\
 & & & \bar{1} & \leftrightarrow & \bar{1} & \downarrow & & \\
 & & & & & \bar{1} & \leftrightarrow & \bar{1} & \\
\hline
 & & 1 & 1 & 0 & 0. & 0 & 1 & 1 \\
 & & & & \bar{1} & \leftrightarrow & \bar{1} & & \\
 & & & 1 & \leftrightarrow & 1 & \downarrow & & \\
 & & & \downarrow & & 1 & \leftrightarrow & 1 & \\
 & & \bar{1} & \leftrightarrow & \bar{1} & & \bar{1} & \leftrightarrow & \bar{1} \\
\hline
 & & 0 & \bar{1} & 1 & \bar{1}. & 1 & \bar{1} & 0 \\
 & & & & 1 & \leftrightarrow & 1 & & \\
 & & & \bar{1} & \leftrightarrow & \bar{1} & \downarrow & & \\
 & & & \downarrow & & \bar{1} & \leftrightarrow & \bar{1} & \\
 & & 1 & \leftrightarrow & 1 & & 1 & \leftrightarrow & 1 \\
\end{array}
$$

As it follows from this example, we have found a special kind of summation (subtraction) called a "swing" phenomenon [53]. If we continue the summation process in the above example, then, by starting from a particular step, the process of the carryover formation turns out to be repetitive and hence the summation becomes *infinite*. The "swing" phenomenon is a variety of the "rapid motion" (races), which arise in digit automata when elements are switching over.

To eliminate the "*swing*" *phenomenon*, we can use the following simple and effective "technical" method [53]. The "*swing*" *phenomenon* arises in the ternary mirror-symmetrical summator (subtractor) in Fig. 5.2, when the carryovers come simultaneously on the some summation step from the two adjacent single-digit summators (subtractors) with the *odd* indices $k$ $(k = 0, \pm2, \pm4, \ldots)$. Then at the second step of the ternary mirror-symmetrical summation (subtraction), the carryovers, formed at the first step, are summarized with the corresponding ternary numerals of the *odd* digits of the summarized numbers. Thanks to such approach, the "*swing*" *phenomenon* is eliminated.

Let's now demonstrate the above method of the elimination of the "*swing*" *phenomenon* at the summation of the "golden" ternary mirror-symmetrical numbers $5 + 5$:

$$
\begin{array}{rccccccc}
k = & 3 & 2 & 1 & 0 & \bar{1} & \bar{2} & \bar{3} \\
5 = & 0 & 1 & \bar{1} & 1. & \bar{1} & 1 & 0 \\
5 = & 0 & 1 & \bar{1} & 1. & \bar{1} & 1 & 0 \\
\hline
s_1 = & & & \bar{1} & \bar{1}. & & \bar{1} & \\
& & & \downarrow & \downarrow & & \downarrow & \\
c_1^1 & & & \downarrow & 1 & \leftrightarrow & 1 & \downarrow \\
c_1^2 & & 1 & \leftrightarrow & 1 & & 1 & \leftrightarrow & 1 \\
\hline
10 = & 1 & \bar{1} & 0 & \bar{1}. & 0 & \bar{1} & 1
\end{array}
$$

The first step of the "*golden*" *ternary mirror-symmetrical summation* involves summarizing all the input "*golden*" *ternary mirror-symmetrical numerals* with the *even* indices $(2, 0, \bar{2})$. The ternary numerals of all digits with the *odd* indices $(3, 1, \bar{1}, \bar{3})$ don't use the first step. The second step involves the summation of all the carryovers, which arise at the first step, with the input ternary numerals of the digits with the *odd* indices. It is important to emphasize that the summation result

$$10 = 1\bar{1}0\bar{1}.0\bar{1}1 \tag{5.39}$$

is represented in the mirror-symmetrical form.

The analysis of all the above examples of the ternary mirror-symmetrical summation (subtraction) shows that both the final summation (subtraction) results and all the intermediate summation (subtraction) results are the *"golden" ternary mirror-symmetrical numbers*, that is, the property of *the mirror symmetry is the invariant about the mirror-symmetrical summation (subtraction)*. This means that mirror-symmetrical *summation (subtraction)* possesses the important mathematical property of the *mirror symmetry*, which can be used for checking the "golden" ternary mirror-symmetrical summator (subtractor) in Fig. 5.2.

## 5.6. The Ternary Mirror-Symmetrical Multiplication and Division

### 5.6.1. Mirror-symmetrical multiplication

The following trivial identity for the golden ratio powers underlies the mirror-symmetrical multiplication:

$$\Phi^{2n} \times \Phi^{2m} = \Phi^{2(n+m)}. \tag{5.40}$$

The rule of the mirror-symmetrical multiplication of the *two single-digit "golden" ternary mirror-symmetrical numbers* is given in Table 5.4.

The ternary mirror-symmetrical multiplication is fulfilled in the *"direct"* code. For technical realization of the *"golden" ternary mirror-symmetrical multiplication* the same "golden" ternary single-digit multiplier as the multiplier for the well-known ternary-symmetrical multiplication described above can be used.

The general algorithm of the multiplication of the two *"golden" ternary multi-digit mirror-symmetrical numbers* reduces to the formation of the partial products in accordance with Table 5.4 and their summation (subtraction) in accordance with the rule

Table 5.4. Ternary mirror-symmetrical multiplication.

| $a_k/b_k$ | $\bar{1}$ | 0 | 1 |
|---|---|---|---|
| $\bar{1}$ | 1 | 0 | $\bar{1}$ |
| 0 | 0 | 0 | 0 |
| 1 | $\bar{1}$ | 0 | 1 |

of the *"golden" ternary mirror-symmetrical summation–subtraction* (Table 5.3).

**Example 5.3.** Multiply the negative ternary mirror-symmetrical number $-6 = \bar{1}01.0\bar{1}$ on the positive ternary mirror-symmetrical number $2 = 1\bar{1}.1$:

$$
\begin{array}{rrrrrrr}
\bar{1} & 0 & 1. & 0 & \bar{1} & & \\
 & & 1 & \bar{1}. & 1 & & \\
\hline
 & \bar{1} & 0. & 1 & 0 & \bar{1} & \\
 & 1 & 0 & \bar{1}. & 0 & 1 & \\
\bar{1} & 0 & 1 & 0. & \bar{1} & & \\
\hline
\bar{1} & 1 & 0 & \bar{1}. & 0 & 1 & \bar{1} \\
\end{array}
$$

The multiplication result in Example 5.3 is formed as the sum of the three partial products. The first partial product $\bar{1}0.10\bar{1}$ is the result of the multiplication of the *"golden" ternary mirror-symmetrical multiplier* $-6 = \bar{1}01.0\bar{1}$ by the minor positive numeral 1 of the *"golden" ternary mirror-symmetrical multiplier* $2 = 1\bar{1}.1$, the second partial product $10\bar{1}.011$ is the result of the multiplication of the same number $-6 = \bar{1}01.0\bar{1}$ on the middle negative numeral $\bar{1}$ of the ternary mirror-symmetrical number $2 = 1\bar{1}.1$, and, finally, the third partial product $\bar{1}010.\bar{1}$ is the result of the multiplication of the same *negative "golden" ternary mirror-symmetrical number* $-6 = \bar{1}01.0\bar{1}$ on the senior positive numeral 1 of the "golden" ternary mirror-symmetrical number $2 = 1\bar{1}.1$.

Note that the product $-12 = \bar{1}10\bar{1}.01\bar{1}$ is represented in the mirror-symmetrical form! Since its higher digit is a negative numeral $\bar{1}$, it follows from this that the product $(-12)$ is the negative "golden" ternary mirror-symmetrical number.

## 5.6.2. Mirror-symmetrical division

The *"golden" ternary mirror-symmetrical division* is carried out in accordance with the division rule of the classic ternary-symmetrical numeral system described above. The general algorithm of the "golden" ternary mirror-symmetrical division reduces to the sequential subtraction of the shifted divisor, which is multiplied with the next ternary numeral of the quotient.

### 5.6.3. The main arithmetical advantages of the ternary mirror-symmetrical arithmetic

We can point out a number of important advantages of the *"golden"* *ternary mirror-symmetrical arithmetic* from a "technical" point of view:

(1) The *"golden" ternary mirror-symmetrical subtraction* is the same arithmetic operation as the *"golden" ternary mirror-symmetrical summation*. The "golden" ternary mirror-symmetrical summation (subtraction) is fulfilled by means of one and the same mirror-symmetrical summator (subtractor) in Fig. 5.2 in the *"direct"* code, that is, without the use of the notions of the *inverse* and *additional* codes.

(2) The sign of the result of the *"golden" ternary mirror-symmetrical summation (subtraction)* is defined automatically because it coincides with the sign of the senior significant ternary numeral of the *"golden" ternary mirror-symmetric representation* of the summation (subtraction) result. The summation (subtraction) results are always represented in the *mirror-symmetrical form*, which allows checking the process of the *"golden" ternary mirror-symmetrical summation (subtraction) according to the property of the "mirror-symmetry"*.

(3) The *"golden" ternary mirror-symmetrical multiplication* reduces to the *"golden" ternary mirror-symmetrical summation (subtraction)*. The *"golden" ternary mirror-symmetrical multiplication* can be carried out in the *"direct"* code, that is, without the use of the notions of the *inverse* and *additional* codes.

(4) The sign of the result of the *"golden" ternary mirror-symmetrical multiplication* is defined automatically because it coincides with the sign of the senior significant ternary numeral of the "golden" ternary mirror-symmetrical result of multiplication. The results of the mirror-symmetrical multiplication are always represented

in the mirror-symmetrical form that allows checking the process of the "golden" ternary mirror-symmetrical multiplication.

(5) The operation of the "*golden*" *ternary mirror-symmetrical division* is a more complicated arithmetical operation compared with the arithmetical operations of the "golden" ternary mirror-symmetrical summation, subtraction and multiplication. For its complexity, the operation of the ternary mirror-symmetrical division is comparable to the same operation in the ternary-symmetrical numeral system used in the ternary computer "Setun" (Moscow University).

## 5.7. Technical Realizations of the Ternary Mirror-Symmetrical Arithmetical Devices

### 5.7.1. General information

By comparing *the classical ternary-symmetrical arithmetic*, based on the following property of the ternary numbers,

$$1 + 1 = 3^n + 3^n = 3^{n+1} - 3^n = 1\bar{1}, \qquad (5.41)$$

with the "*golden*" *ternary mirror-symmetrical arithmetic*, based on the following property of the *golden ratio*,

$$1 + 1 = \Phi^{2n} + \Phi^{2n} = \Phi^{2(n+1)} - \Phi^{2n} + \Phi^{2(n-1)} = 1\bar{1}1, \qquad (5.42)$$

we can see that there is a similarity between (5.41) and (5.42) from the point of view of technical realization. As it is known, the rule of the intermediate sum and carryover formation at the summation (subtraction) of the two single-digit ternary-symmetrical numbers is based on the relation (5.41). This rule consists in the following. The *intermediate sum* $\bar{1}$ and the *carryover* 1, which is spread to the adjacent senior digit are formed.

The identity (5.42) gives the rule of the formation of the intermediate sum and the carryover at the summation (subtraction) of the two single-digit ternary numerals $\{\bar{1}, 0, 1\}$. In accordance with (5.42), the intermediate sum $\bar{1}$ and the carryover 1, which is spread to the left-hand and the right-hand adjacent digits relative to the initial digit, are formed. We can see that *the rules of the formation of the intermediate sum and the carryover at the summation–subtraction of the single-digit ternary-symmetrical and the "golden" ternary mirror-symmetrical numerals coincide.* The distinction consists only in spreading of the carryovers. For the case (5.41), the carryover spreads to the left, that is, to the side of the senior digit; for the case (5.42), the carryover spreads symmetrically relative to the initial digit, that is, to the left and to the right digits simultaneously.

A very important technical conclusion follows from this consideration: *the logic circuits for the realization of the single-digit transformation of the ternary-symmetrical arithmetic and the "golden" ternary mirror-symmetrical arithmetic are identical.* In particular, we can use without any change at the technical realization of the "golden" ternary mirror-symmetrical arithmetic the same logic circuits of the ternary-symmetrical arithmetic (see above).

### 5.7.2. The ternary mirror-symmetrical accumulator

In the previous section, we have developed the multi-digit "golden" ternary mirror-symmetrical summator (subtractor) (Fig. 5.2). This summator (subtractor) is the basis for the *"golden" ternary mirror-symmetrical accumulator*, which is a *"key"* device of the "golden" ternary mirror-symmetrical processor (Fig. 5.3).

The accumulator in Fig. 5.3 has a traditional structure and consists of the multi-digit "golden" ternary mirror-symmetrical

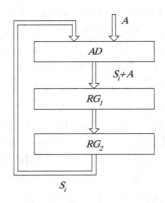

Fig. 5.3. Ternary mirror-symmetrical accumulator.

summator (subtractor) $AD$, the intermediate ternary register $RG_1$, which memorizes the sum $S_1 + A$, arising at the $AD$-output, and the accumulating ternary register $RG_2$.

The *"golden" ternary mirror-symmetrical accumulator* in Fig. 5.3 is a universal device of the *"golden" ternary mirror-symmetrical processor* and underlies the other "golden" ternary mirror-symmetrical devices. By using the additional devices, we can construct the following *"golden" ternary mirror-symmetrical devices*:

(a) If we add to the input $A$ the positive (1) or negative ($\bar{1}$) ternary numerals sequentially, we turn the accumulator into summing or subtracting *"golden" ternary mirror-symmetrical counter.*

(b) If we add the device for the formation of the partial products $A \times b_i \Phi^{2i}$ before the input $A$, we turn the accumulator into the multiplier.

(c) Because the "golden" ternary mirror-symmetrical division reduces to the shift of the divisor and its subtraction from the dividend, then the "golden" ternary mirror-symmetrical accumulator in Fig. 5.2 can be used for designing *the "golden" ternary mirror-symmetrical divider.*

## 5.8. Matrix and Pipeline Mirror-Symmetrical Arithmetical Unit

### 5.8.1. Matrix mirror-symmetrical summator (subtractor)

It is well known that the digital signal processors put forward high demands to the speed of arithmetical devices. The different special structures (matrix, pipeline, etc.) were elaborated for this purpose. We show in this section that the *"golden" ternary mirror-symmetrical arithmetic* contains in itself the interesting possibilities for designing the *fast "golden" ternary arithmetical processors for special applications.*

Let's now consider the many-digit "golden" ternary mirror-symmetrical matrix summator (subtractor) (Fig. 5.4). Each cell of the *"golden" ternary mirror-symmetrical matrix summator*

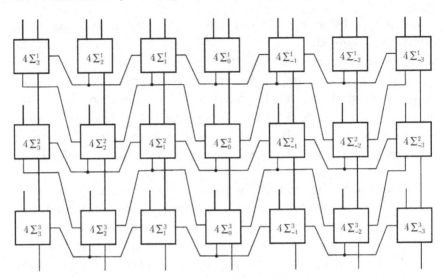

Fig. 5.4. Matrix ternary mirror-symmetrical summator (subtractor).

(*subtractor*) in Fig. 5.4 is a single-digit "golden" ternary mirror-symmetrical full summator (subtractor), which have four inputs and two outputs (see Fig. 5.1(b)).

The matrix seven-digit *"golden" ternary mirror-symmetrical summator (subtractor)* in Fig. 5.4 consists of the 21 single-digit full summators (subtractors), which are arranged in the form of the $7 \times 3$-matrix. Each *"golden" ternary single-digit summator (subtractor)* has a designation $4\Sigma_i^k$, where number 4 means that the summator (subtractor) $4\Sigma_i^k$ has four ternary inputs, the indices $i$ and $k$ in the summator (subtractor) $4\Sigma_i^k$ mean that the summator (subtractor) $4\Sigma_i^k$ refers to the $i$th digit of the ternary mirror-symmetrical code (5.10) and the summator (subtractor) is placed in the $k$th row of the matrix summator (subtractor) in Fig. 5.4.

The inputs of the single-digit summator (subtractor)

$$4\Sigma_3^1, 4\Sigma_2^1, 4\Sigma_1^1, \Sigma_0^1, 4\Sigma_{-1}^1, 4\Sigma_{-2}^1, 4\Sigma_{-3}^1$$

of the first row form the seven-digit input of the *matrix "golden" ternary mirror-symmetrical summator (subtractor)* in Fig. 5.4. The output of the intermediate sum of each single-digit summator (subtractor) is connected to the corresponding input of the next single-digit summator (subtractor) of the same column.

The outputs of the intermediate sum of the single-digit summators (subtractors)

$$4\Sigma_3^3, 4\Sigma_2^3, 4\Sigma_1^3, \Sigma_0^3, 4\Sigma_{-1}^3, 4\Sigma_{-2}^3, 4\Sigma_{-3}^3$$

of the last row form the multi-digit output of the matrix mirror-symmetric summator (subtractor).

The main peculiarity of the matrix mirror-symmetrical summator (subtractor) in Fig. 5.4 consists in a special organization of the connections between the carryover outputs of the single-digit summators (subtractors) and the inputs of the adjacent single-digit summators (subtractors). The carryover outputs of all single-digit summators (subtractors) with the *even* lower indices $(2, 0, \bar{2})$ are connected to the corresponding inputs of the adjacent single-digit summators (subtractors), which are placed in the same row, but the carryover outputs of all the single-digit summators (subtractors) with the *odd* lower indices $(3, 1, \bar{1}, \bar{3})$ are connected with the corresponding inputs of the adjacent single-digit summators (subtractors), which are placed in the lower row. Note that such organization of the carryover connections allows elimination of the above "swing" phenomenon.

Let's consider the operation of the matrix mirror-symmetrical summator (subtractor) on the example of the summation of two equal "golden" ternary mirror-symmetrical numbers:

$$A = 0\ 1\ 1\ 1.1\ 1\ 0 \quad \text{and} \quad B = 0\ 1\ 1\ 1.1\ 1\ 0.$$

The summation is carried out in two stages. Each stage is carried out by means of one row of the single-digit summators (subtractors) and consists of two stages.

**The first stage:** In accordance with Fig. 5.4, the *first step* of the first stage consists in the following. The single-digit summators (subtractors) of the first row with the *even* lower indices $(4\Sigma_2^1, 4\Sigma_0^1, 4\Sigma_2^1)$ form the intermediate sums, which enter the inputs of the second row of summators (subtractors), and the carryovers, which enter the corresponding inputs of the single-digit summators (subtractors) with the *odd* lower indices of the first row ($4\Sigma_3^1$, $4\Sigma_1^1$, $4\Sigma_{\bar{1}}^1$, $4\Sigma_{\bar{3}}^1$). Such transformation of the code information can be represented as follows:

$$
\begin{array}{ccccccc}
0 & 1 & 1 & 1. & 1 & 1 & 0 \\
0 & 1 & 1 & 1. & 1 & 1 & 0 \\
\hline
 & \bar{1} & & \bar{1} & & \bar{1} & \\
\end{array}
$$

$$
\downarrow \quad 1 \;\leftrightarrow\; 1 \;\;\downarrow
$$

$$
1 \;\leftrightarrow\; 1 \qquad\qquad 1 \;\leftrightarrow\; 1
$$

Hence, the first step involves the formation of the intermediate sums and the carryovers on the outputs of the single-digit summators (subtractors) with the *even* lower indices ($2, 0, \bar{2}$).

At the *second step* of the first stage, the single-digit summators (subtractors) with the *odd* lower indices ($3, 1, \bar{1}, \bar{3}$) go into action. In accordance with the entered carryovers, they form the intermediate sums and the carryovers, which enter the single-digit summators (subtractors) of the lower row, that is,

$$
\begin{array}{ccccccc}
0 & 1 & 1 & 1. & 1 & 1 & 0 \\
0 & 1 & 1 & 1. & 1 & 1 & 0 \\
\hline
 & \bar{1} & & \bar{1} & & \bar{1} & \\
\end{array}
$$

$$
\downarrow \quad 1 \;\leftrightarrow\; 1 \;\;\downarrow
$$

$$
1 \;\leftrightarrow\; 1 \qquad\qquad 1 \;\leftrightarrow\; 1
$$

$$
\begin{array}{ccccccc}
\hline
1 & \bar{1} & \bar{1} & 1. & \bar{1} & \bar{1} & 1 \\
\end{array}
$$

$$
1 \;\leftrightarrow\; 1 \;\;\downarrow
$$

$$
1 \;\leftrightarrow\; 1
$$

The first stage is over. We can see that the results of the first stage include some intermediate sum and some carryovers, which enter the summators (subtractors) of the lower row.

**The second stage:** The single-digit summators (subtractors) of the second row with the *even* lower indices ($4\Sigma_2^2$, $4\Sigma_0^2$, $4\Sigma_{\bar{2}}^2$) form the intermediate sums, entering the corresponding inputs of the lower row summators (subtractors) and the carryovers, entering the corresponding inputs of the same row summators (subtractors) with the *odd* lower indices $(4\Sigma_3^2, 4\Sigma_1^2, 4\Sigma_1^2, 4\Sigma_3^2)$, that is,

$$
\begin{array}{ccccccc}
1 & \bar{1} & \bar{1} & \bar{1}. & \bar{1} & \bar{1} & 1 \\
 & 1 & \leftrightarrow & 1 & \downarrow & & \\
 & & & 1 & \leftrightarrow & 1 & \\
\hline
1 & 0 & \bar{1} & 1. & \bar{1} & 0 & 1
\end{array}
$$

Since all the carryovers, which are formed at this stage, turn out to be equal to 0, then the summation (subtraction) process is over at the second stage (this is true only for the considered case). The obtained sum enters the inputs of the lower row summators (subtractors) $4\Sigma_3^3 - 4\Sigma_3^3$ and then appears on the output of the matrix summators (subtractors).

### 5.8.2. The pipeline mirror-symmetrical summator (subtractor)

There are two ways for the extension of the functional possibilities of the "golden" matrix mirror-symmetrical summator (subtractor) in Fig. 5.4. If we place the ternary registers, which consist of the flip-flop-flaps (see above) between the adjacent rows of the single-digit summators (subtractors), then the above "golden" *matrix mirror-symmetrical summator (subtractor)* turns into the *pipeline "golden" ternary mirror-symmetrical summator (subtractor)*. In fact, the code information from the preceding rows of the single-digit summators (subtractors) is memorized in the corresponding ternary registers and the preceding row of the summators (subtractors) becomes ready for further processing. Then, the summators (subtractors) of the lower row process the code information, which enters the lower row of the single-digit summators (subtractors), and simultaneously the top row of the single-digit summators (subtractors) starts processing the new input code information. This means that since the given moment, we get the sums of the numbers $A_1 + B_1, A_2 + B_2, \ldots, A_n + B_n$, which enter the summator (subtractor) during the time period $2\Delta\tau$, where $\Delta\tau$ is the delay time of the single-digit summator (subtractor).

### 5.8.3. The pipeline ternary mirror-symmetrical multiplier

The other way to extend functional possibilities of the pipeline summator (subtractor) consists in the following. We can see in Fig. 5.4 that each single-digit summator (subtractor) of the lower rows has the *free* input. We can use these inputs as new multi-digit inputs of the pipeline summator (subtractor). By using these multi-digit inputs, we can turn the pipeline summator (subtractor) into the *pipeline multiplier*. In this case, the mirror-symmetric multiplication of the two *"golden" ternary mirror-symmetrical numbers* $A(1) \times B(1)$ is carried out in the following manner. The first row of the single-digit summator (subtractor) summarizes the first two partial products $P_1^1 + P_2^1$. This code information enters the second row of the single-digit summators (subtractors). If we send the third partial product $P_3^1$ to the *free* inputs of the second row, we get the sum $P_1^1 + P_2^1 + P_3^1$ on the outputs of the second row. In this case, the first row starts to sum the first two partial products of the next pair of multiplied numbers $A(2) \times B(2)$. The *free* inputs of the third row are used to accept the next partial product $P_4^1$ of the first pair of the multiplied numbers $A(1) \times B(1)$, etc. We can see that the pipeline summator (subtractor) in Fig. 5.4 allows multiplication of many *"golden" ternary mirror-symmetrical numbers* in the pipeline regime. In this connection, the multiplication speed is determined by the time $2\Delta\tau$, where $\Delta\tau$ is the delay time of the single-digit summator (subtractor).

### 5.9. Evaluation of the Ternary Mirror-Symmetrical Arithmetic

### 5.9.1. The ternary mirror-symmetrical arithmetic as the synthesis of Bergman's system and Brousentsov's ternary principle

The *Bergman system* [54] is the first modern mathematical achievement, which underlies the *"golden" ternary mirror-symmetrical arithmetic*. The *Brousentsov ternary principle*, embodied in the ternary computer "Setun" (Moscow University, 1957, Principal Designer Nikolay Brusentsov), is the second original achievement in modern computer science, which can be used in the *"golden"* ternary *mirror-symmetrical arithmetic.*

The *"golden"* ternary *mirror-symmetrical numeral system* with *the irrational base* $\Phi^2 = \frac{3+\sqrt{5}}{2}$ (the square of the golden ratio) [72] retains all the major advantages of the *Bergman system* and the *classical ternary-symmetrical numeral system.* Its main advantage in comparison with conventional ternary-symmetrical numeral system is the original way of checking all the major informational transformations according to the *"principle of mirror symmetry"*. *Therefore, we believe that the problem of implementation of the ternary mirror-symmetrical arithmetic in modern computer technology is a matter of not-so-distant future.*

## 5.9.2. Support of Prof. Donald Knuth

The difficulties associated with the recognition of new ideas and concepts in science and mathematics are well known. The fate of the non-Euclidean geometry, developed by young Russian mathematician Nikolai Lobachevsky (Kazan University), is a classic example. This outstanding mathematical discovery was evaluated very negatively by the Russian leader of mathematical community, academician M.V. Ostrogradsky. But thanks to the support of the outstanding German mathematician Friedrich Gauss, Lobachevsky's geometry was recognized by the international mathematical community, and later the Russian mathematical community was forced to recognize Lobachevsky's geometry as the outstanding mathematical achievement of the 19th century [102].

In this regard, it is appropriate to evaluate the role of the prominent American mathematician and world expert in computer science Prof. Donald Knuth in the development of the numeral systems with irrational bases (Fig. 5.5).

Unfortunately, the mathematicians and computer experts of the 20th century were unable to appreaciate *Bergman's system* as an outstanding mathematical discovery. Prof. Donald Knuth was the only exception. He posted the link to Bergman's article [54] in his best-selling book *The Art of Computer Programming* [123] and in this manner, he supported Bergman's mathematical discovery. Thanks

Fig. 5.5. American mathematician and world expert in computer science, Prof. Donald Knuth.

to Donald Knuth's book [123], Alexey Stakhov learned Bergman's article [54], which influenced Stakhov's research in the theory of numeral systems with irrational bases and their applications.

In 2002, the well-known international magazine *The Computer Journal* (British Computer Society) published Stakhov's article "Brousentsov's ternary principle, Bergman's number system and ternary mirror-symmetrical arithmetic" [72]. Prof. Donald Knuth was the first prominent scientist, who responded to this publication. In his letter to Alexey Stakhov, Prof. Knuth highly appreciated the article [72] and told about his intention to give a description of the "golden" ternary mirror-symmetrical arithmetic in the new edition of his book [123]. This letter of the world famous scientist is the highest award for Stakhov's researches in the field of the new computer arithmetics.

### 5.9.3. Support of Nikolay Brusentsov

Brusentsov's works on the creation of the ternary computer "Setun" the first in the history of science became the main motivation for Alexey Stakhov for the creation of the new "golden" ternary mirror-symmetrical arithmetic based on the *golden ratio*.

On May 29, 2003, Alexey Stakhov made a speech at the meeting of the prestigious scientific seminar "Geometry and Physics"

Fig. 5.6. Nikolay Brusentsov and Alexey Stakhov (Moscow University, May 29, 2003).

(Seminar leader, Prof. Yuri Mikhailov, Department of Theoretical Physics, Moscow University). During the speech, Alexey Stakhov met with the patriarch of the Soviet computer science, Nikolay Brusentsov, the Principal Designer of the ternary computer "Setun" (Fig. 5.6).

Nikolay Brusentsov participated in the discussion of Stakhov's speech and stated as follows:

> *"Fibonacci series and its generalizations, the golden ratio and the following from it the binary $\Phi$-code with irrational base $\Phi = \frac{1+\sqrt{5}}{2}$ and the ternary mirror-symmetrical code with irrational base $\Phi = \frac{3+\sqrt{5}}{2}$ are fundamental components of the structure of the world order. The impressive Stakhov's results will contribute to overcoming the crisis in the foundations of modern science."*

# Chapter 6

# Fibonacci $p$-Codes and Fibonacci Arithmetic for Mission-Critical Applications

## 6.1. "Trojan Horse" of Modern Computers for Mission-Critical Applications

### 6.1.1. The main disadvantage of the binary system

As is known, the binary system was introduced into computer science by John von Neumann (1903–1957) (together with von Neumann's colleagues at the Princeton Institute for Advanced Study) in 1946. The justification of the use of the binary system in electronic computers was one of the most important *"von Neumann's principles"* [97]. At that time, this proposal was an absolutely true decision because the binary system best corresponded to the binary character of the electronic elements and Boolean logic. Moreover, it should take into consideration the fact that at that time other alternative numeral systems in computer science simply did not exist. The choice was very small: *decimal, ternary,* or *binary* systems. Preference was given to the binary system for the case of electronic computers.

However, along with the binary system, its *"Trojan horse"* (according to the apt saying of Russian academician Hetagurov) was introduced into computer technology: *zero redundancy* of the *binary system.* The *zero redundancy* means that all the *binary codewords* are "allowed", which makes impossible the detection of any errors that inevitably (with more or less likely) may occur in electronic systems under the influence of the various external and internal factors (radiation, cosmic rays, electromagnetic connection and so on).

Thus, the next not very optimistic conclusion follows from this consideration. The mankind became hostage to the classical binary system, which is the basis of modern micro-processors and information technology. Therefore, further development of micro-processors and information technology, based on the *binary system*, should be declared inadmissible for certain areas of applications: the *mission-critical applications.*

The binary system cannot be the informational and arithmetical basis for specialized computing and measuring systems (space system, fast transport, complex technological objects, nuclear energy systems, medical systems, military systems and so on), as well as nano-electronic systems, for which the problems of reliability and survivability of computing and measuring systems are very important and come to the fore. Unfortunately, micro-electronics was forced to adopt all the technical solutions of classical computer technology along with the *binary system.* In this field, the problem of the *"Trojan horse"* of the *binary system* (*zero redundancy*) became very important for the micro-processors and the micro-controllers. Currently, the binary system together with its "Trojan horse" begins to take its firm positions in nano-electronics [99, 100], which can lead to unpredictable consequences for further development of information technology.

### 6.1.2. The first Stakhov publications on the redundant numeral systems

For the first time, the studies on the redundant methods of the positional representation of numbers were performed by Alexey Stakhov in the early 1970s in the Taganrog Radio Engineering Institute (Russia) (1971–1977) [55]. At the same time, Stakhov's articles on the redundant positional numeral systems (Fibonacci *p*-codes and codes of the golden *p*-proportions) were published by the famous Russian, English and Ukrainian journals and publishers [6, 16, 17, 19, 46, 53, 55, 56, 58–62, 65, 71, 72, 76, 84, 94, 97, 99, 100].

### 6.1.3. Stakhov's scientific trip to Austria, Stakhov's speech on the joint meeting of Austrian Cybernetics and Computer Societies and the Fibonacci patents by Alexey Stakhov

In 1976, Alexey Stakhov worked for 2 months as the visiting professor of the Vienna University of Technology (Austria). At the final stage of his stay in Austria, Alexey Stakhov made the speech *Algorithmic measurement theory and foundations of computer arithmetic* at the scientific seminar of Graz Technical University and then at the joint meeting of the Austrian Cybernetics and Computer Societies. The speech aroused great interest of the prominent Austrian mathematician Alexander Aigner (Graz Technical University) and other Austrian scientists, who were experts in mathematics and computer science. In this connection, the Soviet Ambassador in Austria, Efremov, wrote a letter to the Soviet State Committee on Science and Technology, which contained the following proposal:

> "Taking into consideration the interest of the Austrian scientists in Prof. Stakhov invention in the field of the new numeral system, based on Fibonacci numbers, it would be appropriate speeding up the process of patenting the invention of Prof. Stakhov abroad what will preserve a priority of Soviet science in this computer field and possibly will give economic effect."

The proposal of the USSR Ambassador in Austria got the approval at the Soviet governmental level and therefore, since 1976, the widespread patenting of Stakhov's inventions on the themes *Fibonacci computers* and *Fibonacci measuring systems* was launched in many countries: USA, Japan, Britain, France, Germany, Canada, and so on.

The main purpose of this patenting was the protection of the priority of the Soviet science in this field. The new computer arithmetic was the subject of patenting. However, in accordance with the patent laws of most countries, it is impossible to get the patent on the mathematical invention, in particular, in the field of *computer arithmetic*, based on the *Fibonacci numbers* and the *golden proportion*.

Therefore, there arose the idea of the indirect protection of the new computer arithmetics by using computer devices, which implemented this arithmetic. *Registers, counters, adders, devices for multiplication and division* and other operating devices of the *Fibonacci computer* were the objects of this patenting. Thus, it was desirable to design such an original operation device, which was a pioneering invention in the field and which would allow the implementation of all other computer devices. As a result of these reasoning, there appeared the idea of the invention of the multicomponent formula, where the first point of the invention was *Fibonacci's pioneering invention*.

What device should be *Fibonacci's pioneering invention*? The device for the reduction of the *Fibonacci code* to the so-called *minimal form* was such a *pioneering invention* (see the following). The invention did not have the analogs and the prototype, and was recognized in the Soviet Union and then in the other countries as a *pioneering invention*. Then, other operating devices, based on the *Fibonacci numbers* (such as counters, adders and so on), were developed on this basis.

What are the results of this unprecedented patenting? On the Soviet inventions in the field of *Fibonacci computers*, more than 60 patents of USA, Japan, England, France, Germany, Canada, Poland, and other countries were given. This shows that the idea of the *Fibonacci computer* is completely new and original, and these patents are the official legal documents, confirming the priority of the Soviet science (and Stakhov's priority) in a new direction in the field of computer science and digital metrology.

The *device for reduction of p-Fibonacci codes to the minimal form* proved to be the pioneering invention from the list of technical solutions submitted for patenting.

In June 1989, according to the initiative of the President of the Ukrainian Academy of Sciences, academician Boris Paton, Stakhov's direction was discussed at a special meeting of the Presidium of the Ukrainian Academy of Sciences; the Presidium members gave a high evaluation to Stakhov's inventions in the Fibonacci area and Stakhov's scientific direction was recognized as a priory direction in Ukraine.

Unfortunately, the so-called *Gorbachev's perestroika*, which brought the collapse of the Soviet Union in 1991, caused an irreparable blow to this scientific direction. Since 1989, the governmental funding of this direction (about 15 million American dollars) was sharply decreased and soon completely stopped; the engineering developments on the Fibonacci topic, which were held in the Vinnitsa Polytechnic Institute in the period 1986–1989, were discontinued.

But this does not mean that the conception of the "Fibonacci computers" was obsolete. On the contrary, at the present stage of the development of computer science and digital metrology, this conception has become even more relevant in modern science, especially in micro-processor and nano-electronic technology. This circumstance became the main motive again to return to these ideas [6, 16, 17, 19, 46, 53, 55, 56, 58–62, 65, 71, 72, 76, 84, 94, 97, 99, 100], which for the first time were described in more detail by Alexey Stakhov in the 1970s to the 1980s in his Russian books [16, 17, 19] and later in his English books [6, 53].

In conclusion, it is important to note that along with the Soviet studies on the *Fibonacci Arithmetic* and *Fibonacci computers*, in the same period, similar studies were performed in the United States (University of Maryland). These studies, which were performed in different countries (Soviet Union, USA, Ukraine, and later in Canada), are a confirmation of the fact that, since the 1970s, the conceptions of the Fibonacci code, Fibonacci arithmetic and Fibonacci computer became widely known in the Soviet, American, Ukrainian and Canadian scientific and technical literature.

## 6.2. Fibonacci $p$-Codes and Their Partial Cases

### 6.2.1. Definition of the Fibonacci $p$-codes

It is proved in [55] that for the given integer $p \in \{0, 1, 2, 3, \ldots\}$, the arbitrary natural number $N$ can be represented as the following sum called *Fibonacci p-code* [55]:

$$N = a_n F_p(n) + a_{n-1} F_p(n-1) + \cdots + a_i F_p(i) + \cdots + a_n F_p(1),$$
$$(6.1)$$

where $a_i \in \{0, 1\}$ is a bit of the $i$th digit of the *Fibonacci p-code* (6.1), $n$ is a number of the bits of the code (6.1), $F_p(i)$ $(i = 1, 2, 3, \ldots, n)$ is the *Fibonacci p-number*, which is the weight of the $i$th digit of the *Fibonacci p-code* (6.1).

In the *Fibonacci p-code* (6.1), the digits weights $F_p(i)$ $(i = 1, 2, 3, \ldots, n)$ are linked by the following recurrent relation:

$$F_p(n) = F_p(n-1) + F_p(n-p-1) \quad \text{for } n > p+1 \qquad (6.2)$$

at the seeds:

$$F_p(1) = F_p(2) = \cdots = F_p(p+1) = 1. \qquad (6.3)$$

The abridged notation of the sum (6.1) is written as follows:

$$N = a_n a_{n-1} \cdots a_i \cdots a_1. \qquad (6.4)$$

The abridged notation (6.4) is called the *Fibonacci p-representation* of the natural number $N$.

### 6.2.2. The partial cases of the Fibonacci $p$-codes

Note that the formula (6.1) specifies the infinite number of the *Fibonacci p-codes* because every $p = 0, 1, 2, 3, \ldots$ "generates" its own *Fibonacci p-code* (6.1).

Let $p = 0$. For this case, the *Fibonacci $(p = 0)$-numbers* $F_0(i)$ coincide with the "binary" numbers, i.e., $F_0(i) = 2^{i-1}$, and therefore, the sum (6.1) reduces to the classical binary system:

$$N = a_n 2^{n-1} + a_{n-1} 2^{n-2} + \cdots + a_i 2^{i-1} + \cdots + a_1 2^0. \qquad (6.5)$$

Let $p = 1$. For this case, Fibonacci ($p = 1$)-numbers $F_1(i)$ coincide with the classical Fibonacci numbers $1, 1, 2, 3, 5, 8, 13, \ldots$, i.e., $F_1(i) = F_i$ and for this case, the sum (6.1) takes the following form, which corresponds to the *classical Fibonacci code* [16, 53]:

$$N = a_n F_n + a_{n-1} F_{n-1} + \cdots + a_i F_i + \cdots + a_1 F_1, \qquad (6.6)$$

where the *Fibonacci weights* $F_i$ ($i = 1, 2, 3, \ldots, n$) are connected by the following relationships:

$$F_i = F_{i-1} + F_{i-2}; \quad F_1 = F_2 = 1. \qquad (6.7)$$

Now, $p = \infty$. For this case, all the *Fibonacci ($p = \infty$)-numbers*, specified by (6.2) and (6.3), are identically equal to 1, i.e., for any $i$, we have $F_p(i) = 1$. For this case, the sum (6.1) takes the following form called the *unitary code*:

$$N = \underbrace{1 + 1 + \cdots + 1}_{N}. \qquad (6.8)$$

Thus, the *Fibonacci p-codes* (6.1) can be considered as a wide generalization of the *binary system* (6.5) ($p = 0$), the *classical Fibonacci code* (6.6) ($p = 1$) and the *unitary code* (6.8) ($p = \infty$). The following numerical-theoretical conclusions follows from this consideration.

(1) The first conclusion follows from the fact that the *Fibonacci p-codes* are the generalization of the *unitary code* (6.8). Note that by its form the *unitary code* (6.8) coincides with the *Euclidean definition of the natural numbers*, which is the beginning of the *elementary theory of numbers*, one of the fundamental theories of ancient mathematics. We can hypothesize from this fact that the *Fibonacci p-codes* (6.1) can be viewed as the *new definitions of the natural numbers*, which implies the idea of the *new number theory*, based on the *Fibonacci p-codes* (6.1). The ancient number

theory, based on the *Euclidean definition* (6.8), is the degenerate case ($p = \infty$) of the more *general theory of numbers* based on the definition (6.1).

(2) The second conclusion follows from the fact that the *Fibonacci p-codes* are the *generalization* of the *binary system* (6.5) ($p = 0$). But we should not forget that the *binary system* (6.6) is the basis of modern computers! But then the following conclusion follows from these facts. Since the *Fibonacci p-codes* (6.1) are the new methods of the *binary* $(0, 1)$ *positional numeral systems*, then we have come to the *new class of the positional numeral systems*, which can lead us to a new class of computers, the *Fibonacci computers*, as a new direction in the development of computer technology! Since the *Fibonacci p-codes* (6.1) are based on the *Fibonacci p-numbers*, which follow from *Pascal's triangle* (the *diagonal sums of Pascal's triangle*), then *the Fibonacci p-codes* (6.1) are the *new unusual mathematical objects*, which unites three fundamental areas of modern science: the *number theory*, which is the basis of mathematics, starting from ancient times, the *combinatorics* [110], which is based on the *binomial coefficients* of *Pascal's triangle*, and finally, *computer science*, which is based on the *binary system* (6.5).

## 6.3.  Unusual Peculiarities of the Fibonacci Representations

### 6.3.1.  A range of number representation in the Fibonacci *p*-codes

Let's consider the set of $n$-digit binary words. A number of them is equal to $2^n$. For the classical binary system (6.5) ($p = 0$), the mapping of the $n$-digit binary words onto the set of natural numbers has the following peculiarities:

(a) **Uniqueness of the mapping:** This means that for the infinite $n$ there is one-to-one correspondence between the natural numbers and the sums (6.1), that is, every *natural number $N$* has the *only* binary representation in the form (6.4).

(b) **Binary system:** For the given $n$, by using the *binary system*
(6.5), we can represent all the natural numbers in the range from
0 to $2^n - 1$, that is, the range of the number representation is
equal to $2^n$. The *minimal* number of 0 and the *maximal* number
of $2^n - 1$ have the following binary representations in the binary
system (6.5):

$$0 = \underbrace{00\ldots0}_{n}; \quad 2^n - 1 = \underbrace{11\ldots1}_{n}. \tag{6.9}$$

For the *Fibonacci p-codes* (6.1), the mapping of the $n$-digit binary
words onto the natural numbers has distinct peculiarities for the case
$p > 0$.

Let $n = 5$. Then for the case $p = 1$, the mapping of the five-digit
*Fibonacci (p = 1)-code* (6.6) onto the natural numbers have the form
represented in Table 6.1.

In Table 6.1, we use the following designations: *FR* means
the *Fibonacci representations*; *FW* means the *Fibonacci weights*.
The analysis of Table 6.1 allows the formulation of the following

Table 6.1. Mapping of the Fibonacci $(p = 1)$-code onto natural numbers.

| $FR/FW$ | 5 | 3 | 2 | 1 | 1 | $N$ | $FR/FW$ | 5 | 3 | 2 | 1 | 1 | $N$ |
|---|---|---|---|---|---|---|---|---|---|---|---|---|---|
| $A_0$ | 0 | 0 | 0 | 0 | 0 | 0 | $A_{16}$ | 1 | 0 | 0 | 0 | 0 | 5 |
| $A_1$ | 0 | 0 | 0 | 0 | 1 | 1 | $A_{17}$ | 1 | 0 | 0 | 0 | 1 | 6 |
| $A_2$ | 0 | 0 | 0 | 1 | 0 | 1 | $A_{18}$ | 1 | 0 | 0 | 1 | 0 | 6 |
| $A_3$ | 0 | 0 | 0 | 1 | 1 | 2 | $A_{19}$ | 1 | 0 | 0 | 1 | 1 | 7 |
| $A_4$ | 0 | 0 | 1 | 0 | 0 | 2 | $A_{20}$ | 1 | 0 | 1 | 0 | 0 | 7 |
| $A_5$ | 0 | 0 | 1 | 0 | 1 | 3 | $A_{21}$ | 1 | 0 | 1 | 0 | 1 | 8 |
| $A_6$ | 0 | 0 | 1 | 1 | 0 | 3 | $A_{22}$ | 1 | 0 | 1 | 1 | 0 | 8 |
| $A_7$ | 0 | 0 | 1 | 1 | 1 | 4 | $A_{23}$ | 1 | 0 | 1 | 1 | 1 | 9 |
| $A_8$ | 0 | 1 | 0 | 0 | 0 | 3 | $A_{24}$ | 1 | 1 | 0 | 0 | 0 | 8 |
| $A_9$ | 0 | 1 | 0 | 0 | 1 | 4 | $A_{25}$ | 1 | 1 | 0 | 0 | 1 | 9 |
| $A_{10}$ | 0 | 1 | 0 | 1 | 0 | 4 | $A_{26}$ | 1 | 1 | 0 | 1 | 1 | 9 |
| $A_{11}$ | 0 | 1 | 0 | 1 | 1 | 5 | $A_{27}$ | 1 | 1 | 0 | 1 | 1 | 10 |
| $A_{12}$ | 0 | 1 | 1 | 0 | 0 | 5 | $A_{28}$ | 1 | 1 | 1 | 0 | 0 | 10 |
| $A_{13}$ | 0 | 1 | 1 | 0 | 1 | 6 | $A_{29}$ | 1 | 1 | 1 | 0 | 1 | 11 |
| $A_{14}$ | 0 | 1 | 1 | 1 | 0 | 6 | $A_{30}$ | 1 | 1 | 1 | 1 | 0 | 11 |
| $A_{15}$ | 0 | 1 | 1 | 1 | 1 | 7 | $A_{31}$ | 1 | 1 | 1 | 1 | 1 | 12 |

peculiarities of the binary representations of the natural numbers in the *Fibonacci p-codes* (6.1) for the case $(p > 0)$. By using the five-digit *Fibonacci* $(p = 1)$-*code* (Table 6.1), we can represent 13 integers in the range 0–12 inclusively. Note that number 13 is the *Fibonacci 1-number* with the index 7, i.e., $F_1(7) = F_7 = 13$. The result of this consideration is the partial case of the following general theorem [55].

**Theorem 6.1.** *For the given integers* $n \geq 0$ *and* $p \geq 0$, *by using the n-digit Fibonacci p-code, we can represent* $F_p(n + p + 1)$ *integers in the range from the minimal number* 0 *to the maximal number* $F_p(n + p) - 1$ *inclusively.*

Note that for the case $p = 0$, $F_0(n + 1) = 2^n$ and Theorem 6.1 reduces to the well-known theorem about the number representation range equal to $2^n$ for the $n$-digit binary system (6.5).

## 6.3.2. Multiplicity of number representation

The *multiplicity* of the *Fibonacci p-representations* of the one and the same natural number $N$ in the form (6.4) is the next peculiarity of the *Fibonacci p-codes* (6.1) for the cases $p > 0$. With the exception of the *minimal* $(N_{\min})$ and *maximal* $N_{\max}$ numbers (see the following)

$$N_{\min} = 0 = \underbrace{00\ldots0}_{n}; \quad N_{\max} = F_p(n + p) - 1 = \underbrace{11\ldots1}_{n}, \quad (6.10)$$

the remaining natural numbers from the range $[N_{\min}, N_{\max}]$ have more than one *Fibonacci p-representation* in the form (6.4). This means that all the natural numbers in the range $[N_{\min}, N_{\max}]$ have *multiple representation* in the *Fibonacci p-codes* (6.1) for the cases $p > 0$.

Let's now consider the mapping of the natural numbers onto the five-digit *Fibonacci representations* (6.5) in accordance with Table 6.2 $(p = 1)$.

Table 6.2. Mapping of the natural numbers onto the
Fibonacci $p$-representations for the case $p = 1$.

| $p = 1$ |
| --- |
| $0 = \{A_0\}$ |
| $1 = \{A_1, A_2\}$ |
| $2 = \{A_3, A_4\}$ |
| $3 = \{A_5, A_6, A_8\}$ |
| $4 = \{A_7, A_9, A_{10}\}$ |
| $5 = \{A_{11}, A_{12}, A_{16}\}$ |
| $6 = \{A_{13}, A_{14}, A_{17}, A_{18}\}$ |
| $7 = \{A_{15}, A_{19}, A_{20}\}$ |
| $8 = \{A_{21}, A_{22}, A_{24}\}$ |
| $9 = \{A_{23}, A_{25}, A_{26}\}$ |
| $10 = \{A_{27}, A_{28}\}$ |
| $11 = \{A_{29}, A_{30}\}$ |
| $12 = \{A_{31}\}$ |

## 6.3.3. "Convolution" and "devolution" of the Fibonacci digits

In Chapter 8, while studying *Bergman's system*, we have introduced
the notions of *convolution* and *devolution*, which allow us to get the
different *"golden" binary representations* of one and the same real
number. These notions are based on the following *golden "summing"*
*relation*:

$$\Phi^i = \Phi^{i-1} + \Phi^{i-2}, \tag{6.11}$$

which links the *"golden" weights* of *Bergman's system* [54].

By comparing *Bergman's system* (4.1) to the classical *Fibonacci
code* (6.6), we find that the recurrent relation (6.7) that connects
the *Fibonacci weights* in the code (6.6) is similar to the *golden
"summing" relation* (6.11) of *Bergman's system* (8.1). This means
that the notions of *convolution* and *devolution* can be applied to the
*classical Fibonacci code* (6.6).

For the general case of $p > 0$, the different *Fibonacci
p-representations* of one and the same integer $N$ in the *Fibonacci*

*p-codes* (6.1) may be obtained from one to another by means of the peculiar code transformations called the *p-convolution* and the *p-devolution* of the binary digits. These code transformations are carried out within the scope of the one binary combination and follow from the basic recurrent relation (6.2), which connects the adjacent *Fibonacci weights* (FW) of the *Fibonacci p-code* (6.1). The idea of such code transformations consists in the following.

Let's consider the *Fibonacci p-representation* (6.4) of the integer $N$ in the *Fibonacci p-code* (6.1). Let's suppose that the binary numerals of the $l$th, $(l-1)$th and $(l-p-1)$th digits in (6.4) are equal to 0, 1, 1, respectively, that is,

$$N = a_n a_{n-1} \cdots a_{l+1} \; \mathbf{01} \; a_{l-2} \cdots a_{l-p} \; \mathbf{1} \; a_{l-p-2} \cdots a_1. \quad (6.12)$$

The *Fibonacci p-representation* (6.12) may be transformed into another *Fibonacci p-representation* of the same number $N$ if, in accordance with the recurrent relation (6.2), we replace the binary numerals $\mathbf{1}$ in the $(l-1)$th and $(l-p-1)$th digits by the binary numeral $\mathbf{0}$ and the binary numeral $\mathbf{0}$ in the $l$th digit by the binary numeral $\mathbf{1}$, that is,

$$N = \begin{cases} a_n & \cdots & a_{l+1} & \mathbf{0} & \mathbf{1} & a_{l-2} & \cdots & a_{l-p} & \mathbf{1} & a_{l-p-2} & \cdots & a_1 \\ & & & \downarrow & \downarrow & & & & \downarrow & & & \\ a_n & \cdots & a_{l+1} & \mathbf{1} & \mathbf{0} & a_{l-2} & \cdots & a_{l-p} & \mathbf{0} & a_{l-p-2} & \cdots & a_1. \end{cases}$$
$$(6.13)$$

Such transformation of the *Fibonacci p-representation* (6.12) is named the *p-convolution* of the $(l-1)$th and $(l-p-1)$th digits into the $l$th digit.

The initial *binary representation* (6.12) can be restored if we carry out the following code transformation over the *Fibonacci p-representation* of the number $N$ given by (6.13):

$$N = \begin{cases} a_n & \cdots & a_{l+1} & \mathbf{1} & \mathbf{0} & a_{l-2} & \cdots & a_{l-p} & \mathbf{0} & a_{l-p-2} & \cdots & a_1 \\ & & & \downarrow & \downarrow & & & & \downarrow & & & \\ a_n & \cdots & a_{l+1} & \mathbf{0} & \mathbf{1} & a_{l-2} & \cdots & a_{l-p} & \mathbf{1} & a_{l-p-2} & \cdots & a_1. \end{cases}$$
$$(6.14)$$

Such transformation of the code representation is named the *p-devolution* of the $l$th digit into the $(l-1)$th and $(l-p-1)$th digits.

Note that in accordance with (6.2), the fulfillment of the *p-convolution* and the *p-devolution* in the *Fibonacci p-representation* (6.12) does not change the initial number $N$ represented by the *Fibonacci p-representation* (6.12).

### 6.3.4. The $p$-convolution and $p$-devolution for the case $p = 1$

For the case $p = 1$, the *p-convolution* and the *p-devolution* take the simplest form for the technical realization:

$$\begin{cases} 011 \rightarrow 100 \ (convolution), \\ 100 \rightarrow 011 \ (devolution). \end{cases} \tag{6.15}$$

Let's now consider the peculiarities of *convolution* and *devolution* for the lowest digits of the *Fibonacci $(p = 1)$-representations*. As is well known for the case $p = 1$, the weights of the two lowest digits of the Fibonacci $(p = 1)$-code is identically equal to 1, that is, $F_1 = F_2 = 1$. Then the operations of "devolution" and "convolution" for these digits are performed as follows:

$$\begin{cases} 01 \rightarrow 10 \ (convolution), \\ 10 \rightarrow 01 \ (devolution). \end{cases} \tag{6.16}$$

### 6.3.5. MINIMAL $p$-FORM for the Fibonacci $p$-code

In Chapter 4, we have introduced the conception of the MINIMAL FORM for the *"golden" representation of real numbers* in *Bergman's system* [54]. The main peculiarity of the *"golden" representation* in the MINIMAL FORM consists in the fact that *in the* MINIMAL FORM for the case $(p = 1)$, the **two bits of 1 together do not meet**.

We take without proof the following theorem proved in [55].

**Theorem 6.2.** *For the given integers $p \geq 0$ and $n \geq p+1$, the arbitrary integer $N$ can be represented in the only form:*

$$N = F_p(n) + R_1, \tag{6.17}$$

*where*

$$0 \leq R_1 < F_p(n-p). \tag{6.18}$$

Note that for the case $p = 0$, we have $F_0(n) = 2^{n-1}$, and therefore, the expressions (6.17) and (6.18) take the following well-known (in the "binary" arithmetic) form:

$$N = 2^{n-1} + R_1, \quad 0 \leq R_1 < 2^{n-1}. \tag{6.19}$$

Let's represent the integer $N$ according to the formulas (6.17) and (6.18) and then let's represent all the remainders $R_1, R_2, R_k$, which can arise in the process of such representation, according to the formulas (6.17) and (6.18). We will continue this process up to obtaining the remainder equal to 0. As a result of this decomposition of number $N$, we obtain the peculiar *Fibonacci p-representation* of the natural number $N$ in the *Fibonacci p-code* (6.1). Its peculiarity is the fact that in the *Fibonacci p-representation* of the *natural number* $N$, specified by (6.4), no less than $p$ bits of 0 follow after every bit $a_l = 1$ from the left to the right, that is,

$$a_{l-1} = a_{l-2} = a_{l-p} = 0.$$

Such a representation of integer $N$ is called the MINIMAL $p$-FORM of the *Fibonacci p-representation* (6.4).

For example, by using the above algorithm, we can obtain the following MINIMAL $p$-FORMS of number 25 for the cases $p = 1$ and $p = 2$ (Table 6.3).

A peculiarity of the binary representations of number 25, given by Table 6.3, consists in the following. For the case $p = 1$, not less than *one* bit of 0 follows after every bit of 1 from the left to the right in the *Fibonacci ($p = 1$)-representation* of number 25; for the case $p = 2$, not less then *two* bits of 0 follow after every bit of 1 from the left to the right in the *Fibonacci ($p = 2$)-representation* of the same number 25.

Table 6.3. MINIMAL FORMS of the Fibonacci 1- and 2-representations.

| $p=1$ | $F_1(i)$ | 55 | 34 | 21 | 13 | 8 | 5 | 3 | 2 | 1 | 1 |
|---|---|---|---|---|---|---|---|---|---|---|---|
| | $25=$ | 0 | 0 | 1 | 0 | 0 | 0 | 1 | 0 | 1 | 0 |
| $p=2$ | $F_2(i)$ | 19 | 13 | 9 | 6 | 4 | 3 | 2 | 1 | 1 | 1 |
| | $25=$ | 1 | 0 | 0 | 1 | 0 | 0 | 0 | 0 | 0 | 0 |

**Corollary from Theorem 6.2.** *For the given p $(p = 0, 1, 2, 3, \ldots)$, every natural number N has the only MINIMAL p-FORM in the Fibonacci p-representation (6.4).*

This means that for the case $n \to \infty$, there is the *one-to-one mapping* of natural numbers onto the MINIMAL p-FORMs of the *Fibonacci p-representations* (6.4).

Note that for the cases $n \to \infty$ and $p = 0$ (the *classical binary system*), there is the *one-to-one* mapping of the natural numbers onto the binary representations in the form (6.4). This means that every binary representation (6.4) is its "MINIMAL FORM" for the case $p = 0$.

We take without proof the following theorem proved in [55].

**Theorem 6.3.** *For the given integer p $\geq 0$ by using the n-digit Fibonacci p-code, in the MINIMAL p-FORM, we can represent $F_p(n+1)$ integers in the range from 0 to $F_p(n+1) - 1$ inclusively.*

### 6.3.6. The base of the Fibonacci p-code

For the case $p = 0$, the base of the binary system (6.5) is calculated as the ratio of the adjacent digit weights, that is,

$$\frac{2^k}{2^{k-1}} = 2.$$

Let's apply this principle to the *Fibonacci p-codes* (6.1) and consider the ratio:

$$\frac{F_p(k)}{F_p(k-1)}. \tag{6.20}$$

A limit of the ratio (6.20) for the case $n \to \infty$ is named the *base of the Fibonacci p-code* (6.1). It is easy to prove [16, 55] that

$$\lim_{k \to \infty} \frac{F_p(k)}{F_p(k-1)} = \Phi_p, \qquad (6.21)$$

where $\Phi_p$ is the *golden p-proportion*, the positive root of the algebraic equation $x^{p+1} - x^p - 1 = 0$ $(p = 0, 1, 2, 3, \ldots, \infty)$.

This means that the base of the *Fibonacci p-codes* (6.1) for the case $p > 0$ is the irrational number $\Phi_p$, that is, for the case $p > 0$, the *Fibonacci p-codes* (6.1) are *numeral systems with irrational bases.*

### 6.3.7. Code redundancy of the Fibonacci *p*-codes

For the case $p = 0$, the *Fibonacci* $(p = 0)$-*code* (the *classical binary code*) is non-redundant. But for the case $p > 0$, all the *Fibonacci p-codes* are redundant, and their *code redundancy* shows itself in *multiplicity* of the *Fibonacci p-representations* of the one and the same natural number $N$. Theorems 6.1 and 6.3 allow the calculation of the *code redundancy* of the *Fibonacci p-codes* for the cases $p > 0$ in comparison with the classical binary system $(p = 0)$, which is a non-redundant binary code.

We can calculate the *relative code redundancy r* by the following formula [16, 55]:

$$r = \frac{n - m}{m} = \frac{n}{m} - 1, \qquad (6.22)$$

where $n$ and $m$ are the code lengths of the redundant and non-redundant codes, respectively. Note that the definition of the *code redundancy*, given by (6.22), characterizes the relative increase of the code length of the redundant code in the comparison to the non-redundant code for the representation of the one and the same number range.

Theorems 6.1 and 6.3 determine the ranges of the number representations in *Fibonacci p-codes* for the two cases: (a) when we use all the possible *Fibonacci p-representations* (Theorem 6.3) and (b) when we use only the MINIMAL FORMS of the *Fibonacci p-representations* (Theorem 6.2). For the case (a), we can represent

$F_p(n+p+1)$ natural numbers, for the case (b), we can only represent $F_p(n+1)$ natural numbers.

For the cases (a) and (b), we need to use either $m_1 \approx \log_2 F_p(n+p+1)$ or $m_2 \approx \log_2 F_p(n+1)$ binary digits of the non-redundant code, respectively. By using (6.21), we can obtain the following general formulas for the calculation of the *relative code redundancy* of the *Fibonacci p-codes* (6.1):

$$r_1 = \frac{n}{\log_2 F_p(n+p+1)} - 1, \tag{6.23}$$

$$r_2 = \frac{n}{\log_2 F_p(n+1)} - 1. \tag{6.24}$$

The simplest redundant *Fibonacci p-code* (6.1) is the code corresponding to the case $p = 1$. We can calculate the limiting value for the *relative redundancy* for this code. For this case, the formulas (6.23) and (6.24) take the following forms, respectively:

$$r_1 = \frac{n}{\log_2 F_{n+2}} - 1, \tag{6.25}$$

$$r_2 = \frac{n}{\log_2 F_{n+1}} - 1, \tag{6.26}$$

where $F_{n+2}$ and $F_{n+1}$ are the *classical Fibonacci numbers*.

We can represent the Fibonacci numbers $F_{n+2}$ and $F_{n+1}$ by using *Binet's formulas*:

$$F_n = \begin{cases} \dfrac{\Phi^n + \Phi^{-n}}{\sqrt{5}} & \text{for } n = 2k+1, \\[3mm] \dfrac{\Phi^n - \Phi^{-n}}{\sqrt{5}} & \text{for } n = 2k. \end{cases} \tag{6.27}$$

For a large $n$, we can write *Binet's formulas* (6.27) in the following approximate form:

$$F_n \approx \frac{\Phi^n}{\sqrt{5}}. \tag{6.28}$$

By using (6.21) and by substituting the approximate values for $F_{n+2}$ and $F_{n+1}$, determined by (6.28),

$$F_{n+2} \approx \frac{\Phi^{n+2}}{\sqrt{5}} \quad \text{and} \quad F_{n+1} \approx \frac{\Phi^{n+1}}{\sqrt{5}}$$

into the formulas (6.25) and (6.26), we get the following formulas:

$$r_1 = \frac{n}{(n+2) \log_2 \Phi - \log_2 \sqrt{5}} - 1, \tag{6.29}$$

$$r_2 = \frac{n}{(n+1) \log_2 \Phi - \log_2 \sqrt{5}} - 1. \tag{6.30}$$

If we direct $n \to \infty$ in the expressions (6.29) and (6.30), we can see that they coincide for the case $n \to \infty$. Here, the limiting value of the *relative code redundancy* for the case $p = 1$ is determined by the following formula:

$$r = \frac{1}{\log_2 \Phi} - 1 = 0.44. \tag{6.31}$$

Thus, the limiting value of the *relative code redundancy* of the Fibonacci ($p = 1$)-code directs in limit to 0.44 (44%).

## 6.4. The Simplest Fibonacci Arithmetical Operations

### 6.4.1. Comparison of numbers in the Fibonacci p-codes

As shown above, *Fibonacci p-codes* (6.1) are a new class of the *positional numeral systems*. The *Fibonacci p-codes* (6.1) are similar to the *binary system* (6.5) and are its generalization. Therefore, all the well-known properties of the *binary system* (6.5) can be used to create the *Fibonacci arithmetic*, although the distinction between them consists in the fact that we should take into consideration the fundamental features of the *Fibonacci p-codes* (6.1) for the case $p > 0$, in particular, the ones such as the *multiplicity* of number representation and MINIMAL FORM.

Let us begin from such an important arithmetical operation as the *comparison of numbers*, which has certain peculiarities for the *Fibonacci p-codes* (6.1) for the case $p > 0$. The comparison of the

*Fibonacci p-representations* $A$ and $B$ is similar to the comparison of numbers in the *binary system*. The main peculiarity consists in the fact that *before the comparison, we should reduce the compared Fibonacci p-representations to the* MINIMAL FORM.

For example, we need to compare the two Fibonacci 1-representations:

$A = 00111101101$ and $B = 00111110110$ for the case of the *Fibonacci 1-code* (6.6).

The comparison of the above numbers is carried out in two steps:

(1) Reduce the compared *Fibonacci 1-representations* to the MINIMAL FORM:

$$A = 01010010010 \quad \text{and} \quad B = 01010100000. \qquad (6.32)$$

(2) Compare digit by digit the MINIMAL FORMS (6.32), by starting from the senior digit until obtaining the first pair of the distinct bits (see in bold):

$$A = 01010\mathbf{0}10010,$$

$$B = 01010\mathbf{1}00000.$$

We can see that the first non-coincident pair of digits (see in bold) in the compared MINIMAL FORMS of the *Fibonacci 1-representations* $A$ and $B$ contains the bit of 0 in the MINIMAL FORM of the first *Fibonacci 1-representation* $A$ and the bit of 1 in the MINIMAL FORM of the second *Fibonacci 1-representation* $B$. This means that $B > A$.

## 6.4.2. The Fibonacci summing and subtracting counters

The algorithm of the *summing Fibonacci counter* consists in the *addition* of the bit 1 to the *least Fibonacci digit* (LFD) according to the principle

$$N' = N + 1. \qquad (6.33)$$

The algorithm includes the two stages:

**The first stage:** Before the addition of the bit 1 to LFD, the initial *Fibonacci p-representation*, corresponding to number $N$, by using the *convolution*, reduces to such a form that the value of LFD will be equal to 0.

**The second stage:** We add the bit of 1 to LFD, which leads to the increase of number in the *Fibonacci counter* to the value of $N + 1$.

Then, we repeat the first and second stages until overflowing *Fibonacci counter*.

Let's demonstrate this algorithm for the following example.

**Example 6.1.** The algorithm of the operation of the Fibonacci summing counter consists of the following:

$$
\begin{aligned}
000000 + 1 &= 0000\,\boxed{01} = 000010 = 1\\
000010 + 1 &= 000\,\boxed{011} = 000100 = 2\\
000100 + 1 &= 0001\,\boxed{01} = 000110 = 3\\
00\,\boxed{011}\,0 + 1 &= 0010\,\boxed{01} = 001010 = 4\\
001010 + 1 &= 001\,\boxed{011} = 001100 = 5\\
0\,\boxed{011}\,00 + 1 &= 0100\,\boxed{01} = 010010 = 6\\
010010 + 1 &= 010\,\boxed{011} = 010100 = 7 \qquad (6.34)\\
010100 + 1 &= 0101\,\boxed{01} = 010110 = 8\\
01\,\boxed{011}\,0 + 1 &= 01\,\boxed{011}\,1 = \boxed{011}\,0\,\boxed{01} = 100010 = 9\\
100010 + 1 &= 100\,\boxed{011} = 100100 = 10\\
100100 + 1 &= 1001\,\boxed{01} = 100110 = 11\\
10\,\boxed{011}\,0 + 1 &= 1010\,\boxed{01} = 101010 = 12\\
101010 + 1 &= 101\,\boxed{011} = 1\,\boxed{011}\,00 = \boxed{110}\,000 = 000000
\end{aligned}
$$

Here, we mark in *bold* those situations, where we can carry out the *convolutions* in the *Fibonacci representations*. Consider, for example, the situation of the transition of the *Fibonacci 1-representation* of the number $8 = 010110$ to the Fibonacci 1-representation of number 9:

$$
01\,\boxed{011}\,0 + 1 = 01\,\boxed{011}\,1 = \boxed{011}\,0\,\boxed{01} = 100010 = 9. \qquad (6.35)
$$

In this case, we add the bit of 1 to LFD of the *Fibonacci 1-representation* of number 8. As a result, after performing all the *convolutions*, we get *Fibonacci 1-representation* of number 9 in the form $9 = 100010$. Note that the bottom row of the Eq. (6.34) corresponds to the overflow of the *Fibonacci counter*.

The algorithm of the *subtraction* of 1's from the *Fibonacci p-code* (6.1) is carried out by the subtraction of the bit of 1 from LFD of the *Fibonacci p-representation* of number $N$ until the value of the LFD becomes equal to 1.

Let's consider the example of the functioning of the *Fibonacci counter* in the *subtracting* regime.

**Example 6.2.** The algorithm of the Fibonacci counter in the subtracting regime for $p = 1$:

$$
\begin{aligned}
1111 - 1 &= 11\boxed{10} = 1101 = 6 \\
1101 - 1 &= 1\boxed{100} = 1011 = 5 \\
1011 - 1 &= 10\boxed{10} = 1001 = 4 \\
\boxed{100}1 - 1 &= 0110 = 0101 = 3 \\
0101 - 1 &= 0\boxed{100} = 0011 = 2 \\
0011 - 1 &= 00\boxed{10} = 0001 = 1 \\
0001 - 1 &= 0000 = 0
\end{aligned}
\tag{6.36}
$$

Here, we mark in *bold* those situations, where we can realize the *devolutions* in the *Fibonacci 1-representations*. Thus, the feature of the *Fibonacci counter* in the subtracting regime consist in the fact that in any situation, the transition from the *Fibonacci 1-representation* of the number $N$ to the *Fibonacci 1-representation* of the number $N - 1$ is performed by using the sequential *devolutions* of the bit 1 in LFD.

The above algorithms of the Fibonacci counters in the *summing* and *subtracting* regimes show that they are the prerequisites for the construction of the high-speed *Fibonacci counters* (without the use of complex schemes of *group transfer*). That is, the simple example shows certain advantages of the Fibonacci code (6.6) in comparison to the *classical binary code* (6.5).

### 6.4.3. Fibonacci summation of the binary numerals

It is well known that the classical *binary summation* is based on the following elementary identity for the *binary numbers*:

$$2^k + 2^k = 2^{k+1}, \qquad (6.37)$$

where $2^k$ and $2^{k+1}$ are the weights of the $k$th and $(k+1)$th digits of the *binary code* (6.5).

We begin the "deduction" of the rule of summation of the *binary numerals* of the *Fibonacci p-representation* from the analysis of the sum:

$$F_p(k) + F_p(k), \qquad (6.38)$$

where $F_p(k)$ is the weight of the $k$th digit of the *Fibonacci p-code* (6.1).

Let $p = 1$. For this case, we have

$$F_{p=1}(k) = F_k, \qquad (6.39)$$

where $F_k$ is the *classical Fibonacci numbers*: $1, 1, 2, 3, 5, 8, 13, \ldots$, given by the recurrent relation (6.7). By using (6.7) and (6.39), we can represent the sum (6.38) (for the case $p = 1$) as follows:

$$\text{(a)} \quad F_k + F_k = F_k + F_{k-1} + F_{k-2}, \qquad (6.40)$$

$$\text{(b)} \quad F_k + F_k = F_{k+1} + F_{k-2}. \qquad (6.41)$$

Table 6.4 demonstrates the principle of the Fibonacci summation $(p = 1)$ of the two numerals $a_k + b_k$ with the equal indexes $k$ based on (6.40) and (6.41).

As it follows from Table 6.4, the rules of the *Fibonacci summation* of the binary numerals for the case $p = 1$ coincide with the *classical binary summation* for the cases: $0 + 0 = 0$, $0 + 1 = 1$,

Table 6.4. Fibonacci summation of the
binary numerals (for the case $p = 1$).

$$0 + 0 = \quad 0$$
$$0 + 1 = \quad 1$$
$$1 + 0 = \quad 1$$
$$1 + 1 = \quad 1\ 1\ 1 \ (a)$$
$$1 + 1 = 1\ 0\ 0\ 1 \ (b)$$

$1 + 0 = 1$. For the case $1 + 1$, the rules of the *Fibonacci summation* of the binary numerals (for the case $p = 1$) are reduced to the following:

**Rule 1 (the case (a)):** At the summation of the numerals of 1's $(1+1)$ in the $k$th digit of the summable *Fibonacci representations* (for $p = 1$), the bit of 1 is inscribed into the $k$th digit of the intermediate sum and there arises the carryover of the two bits of 1 from the $k$th digit at the adjacent $(k-1)$th and $(k-2)$th digits.

**Rule 2 (the case (b)):** The method (b) is based on the identity (6.41) and assumes another rule of the Fibonacci summation of the numerals of 1's $(1 + 1)$ in the $k$th digit of the summable *Fibonacci representations* (for $p = 1$). The bit of 0 is inscribed to the $k$th digit of the intermediate sum and the two carryover of 1 come to the other digits, namely, to the $(k + 1)$th and $(k - 2)$th digits.

### 6.4.4. Fibonacci summation of the multi-bit numbers (for the case $p = 1$)

The summation of the multi-digit numbers in the *Fibonacci code* $(p = 1)$ is carried out in accordance with Table 6.4. However, before the Fibonacci summation, we should adhere to the following rules:

**Rule 3:** Before the summation, the summable *Fibonacci representations* reduce to the MINIMAL FORM.

**Rule 4:** In accordance with Table 6.4, it is necessary to form the multi-bit intermediate sum and the multi-bit carryover for all digits.

**Rule 5:** The multi-bit intermediate sum reduces to the MINIMAL FORM and is then summarized with the multi-bit carryover.

**Rule 6:** The summation process continues in accordance with the Rules 4 and 5 until the obtaining of the multi-bit carryover equal to 0. The last intermediate sum, which reduces to the MINIMAL FORM, is the result of the multi-bit *Fibonacci summation*.

For the above *Fibonacci summation*, we need to add the following additional rule.

**Rule 7:** Let's consider the case where we have the two binary 1's in the $k$th digits of the summable *Fibonacci representations*. It follows from the property of the MINIMAL FORM that the bits of the $(k + 1)$th and $(k - 1)$th digits of both the summable *Fibonacci representations* are always equal to 0. It is clear that for this case, the intermediate sums, arising at the summation of the $(k+1)$th and $(k - 1)$th digits of both the summable *Fibonacci representations*, are always equal to 0. This means that we can place one of the carryovers, arising at the summation of the $k$th *significant digits* $(1 + 1)$, at once to the $(k - 1)$th digit of the intermediate sum (for the (a) method of the *Fibonacci summation*) or to the $(k + 1)$th digit of the intermediate sum (for the (b) method of the *Fibonacci summation*).

### 6.4.5. Fibonacci subtraction

As well known, the method of the "direct" number subtraction in the classical binary arithmetic $(p = 0)$ is based on the following property of the binary numbers:

$$2^{n+k} - 2^n = 2^{n+k-1} + 2^{n+k-2} + \cdots + 2^n. \qquad (6.42)$$

Write now the similar identity for the Fibonacci 1-numbers:

$$F_{n+k} - F_n = F_{n+k-2} + F_{n+k-3} + \cdots + F_{n-1}. \qquad (6.43)$$

Table 6.5. Fibonacci subtraction
table for the case $p = 1$.

| | |
|---|---|
| $0 - 0 =$ | $0$ |
| $1 - 1 =$ | $0$ |
| $1 - 0 =$ | $0\ 1\ 1$ |
| $1\ 0 - 1 =$ | $0\ 1$ |
| $1\ 0\ 0 - 1 =$ | $1\ 1$ |
| $1\ 0\ 0\ 0 - 1 =$ | $1\ 1\ 1$ |

By using the identity (6.43) and Fibonacci recurrent relation (6.5), we can construct the Fibonacci subtraction rule presented in Table 6.7.

The direct Fibonacci subtraction of the multi-digit numbers uses the following rules:

**Rule 1:** Before subtraction, the subtrahend Fibonacci representations reduces to the MINIMAL FORM.

**Rule 2:** The MINIMAL FORMS of the *subtrahend Fibonacci representations* are compared according to the comparison rule of the Fibonacci subtrahend representations and then in accordance with Table 6.5, the smaller Fibonacci representation is subtracted from the bigger Fibonacci representation.

Note that these arithmetical algorithms are new technical solutions. They were developed by Alexey Stakhov in the early 1970s and these algorithms have been used as the basis of comparison, summation and subtraction devices, as well as summing and subtracting Fibonacci counters patented abroad.

There are other rules of arithmetical operations in the Fibonacci code (6.5). In Ref. [6], the rules of Fibonacci subtraction, based on the use of the concepts of *additional* and *inverse* Fibonacci codes, are developed. In the same book [6], the original methods of the Fibonacci summation and subtraction, based on the so-called *basic micro-operations*, are suggested.

## 6.5. Fibonacci Multiplication and Division

### 6.5.1. Fibonacci multiplication

The analysis of the *ancient Egyptian multiplication* [133] leads us to the following method of the *Fibonacci multiplication*.

Consider now the product $P = A \times B$, where the numbers $A$ and $B$ are represented in the Fibonacci $p$-code (6.1). By using the representation of the multiplier $B$ in the Fibonacci $p$-code (6.1), we can write the product $P = A \times B$ in the following form:

$$P = A \times b_n F_p(n) + A \times b_{n-1} F_p(n-1) + \cdots$$
$$+ A \times b_i F_p(i) + \cdots + A \times b_1 F_p(1), \qquad (6.44)$$

where $F_p(i)$ is the Fibonacci $p$-number.

The following algorithm of the *Fibonacci p-multiplication* follows from Eq. (6.44). The multiplication is reduced to the summation of the partial products of the kind $A \times b_i F_p(i)$. They are formed from the multiplier $A$ according to the special procedure that reminds the *ancient Egyptian multiplication* [133].

### 6.5.2. Example of the Fibonacci multiplication

Let's now demonstrate the "Fibonacci multiplication" for the case of Fibonacci 1-code. Let's find the product: $41 \times 305$ by using Fibonacci 1-code (6.6).

A solution of the task consists of the following steps:

1. Construct the table consisting of the three columns: $F, G$ and $P$ (see Table 6.6).
2. Insert the Fibonacci 1-sequence (the classical Fibonacci numbers) 1, 1, 2, 3, 5, 8, 13, 21, 34 to the $F$-column of Table 6.6.
3. Insert the generalized Fibonacci 1-sequence: 305, 305, 610, 915, 1525, 2440, 3965, 6505, 10370, which is formed in the $G$-column from the first multiplier 305, according to the "Fibonacci recurrent relation".

Table 6.6. Fibonacci multiplication for
the case $p = 1$.

| $F$ | $G$ | $P$ |
|---|---|---|
| 1 | 305 | |
| 1 | 305 | |
| /2 | 610 | → 610 |
| 3 | 915 | |
| /5 | 1525 | → 1525 |
| 8 | 2440 | |
| 13 | 3965 | |
| 21 | 6505 | |
| /34 | 10370 | → 10370 |
| 41 = 34 + 5 + 2 | 41 × 305 | = 12505 |

4. Mark by the inclined line (/) and bold all the $F$-numbers that give the second multiplier in the sum $(41 = 34 + 5 + 2)$.
5. Mark by bold all the $G$-numbers corresponding to the marked $F$-numbers and rewrite them to the $P$-column.
6. By summarizing all the $P$-numbers, we obtain the product: $41 \times 305 = 12\,505$.

Fibonacci multiplication algorithm in Table 6.6 is easily generalized for the case of the Fibonacci $p$-codes (6.1) for arbitrary $p$.

### 6.5.3. Example of Fibonacci division

Let's divide the number 481 (the dividend) by the number 13 (the divisor) in the Fibonacci 1-code (6.5) (the case $p = 1$).

The Fibonacci division consists of several stages.

## The first stage:

(1) Construct the table consisting of three columns: $F$, $G$ and $D$ (see Table 6.7).
(2) Insert the Fibonacci 1-sequence (the classical Fibonacci numbers) 1, 1, 2, 3, 5, 8, 13, 21, 34, 55 into the $F$-column of Table 6.7.

Table 6.7. The first stage of the Fibonacci division ($p = 1$).

| $F$ | $G$ | $D$ |
|-----|-----|-----|
| 1 | 13 | $\leq 481$ |
| 1 | 13 | $\leq 481$ |
| 2 | 26 | $\leq 481$ |
| 3 | 39 | $\leq 481$ |
| 5 | 65 | $\leq 481$ |
| 8 | 104 | $\leq 481$ |
| 13 | 169 | $\leq 481$ |
| 21 | 273 | $\leq 481$ |
| /**34** | **442** | $\leq 481$ |
| 55 | 615 | $> 481$ |
| $R_1$ | = | $481 - 442 = 39$ |

(3) Insert the generalized Fibonacci 1-sequence: 13, 13, 26, 39, 65, 104, 169, 273, 442, 615, formed from the divisor 13, according to the "Fibonacci recurrent relation", into the $G$-column.

(4) Compare sequentially every $G$-number with the dividend 481, inscribed into the $D$-column, and fix the result of comparison ($\leq$ or $>$) until we obtain the first comparison result of the kind ($>$): $615 > 481$.

(5) Mark with bold and with the inclined line (/) the Fibonacci number 34, which corresponds to preceding $G$-number 442, and mark this number with bold too. Calculate the difference: $R_1 = 481 - 442 = 39$.

**The second stage:** The second stage of the Fibonacci 1-division is a repetition of the first stage, where we use instead of the dividend 481 the difference $R_1 = 39$ (see Table 6.7).

Since the second difference $R_2 = 39 - 39 = 0$, then the Fibonacci 1-division is over in the second stage. The result of the Fibonacci division is equal to the sum of all the marked $F$-numbers, obtained on all stages (see Tables 6.7 and 6.8):

$$481{:}13 = 34 + 3 = 37.$$

Table 6.8. The second stage of the
Fibonacci division ($p = 1$).

| $F$ | $G$ | $D$ |
|-----|-----|-----|
| 1 | 13 | $\leq 39$ |
| 1 | 13 | $\leq 39$ |
| 2 | 26 | $\leq 39$ |
| /3 | **39** | $\leq 39$ |
| 5 | 65 | $> 39$ |
| $R_2$ | $=$ | $\mathbf{39 - 39 = 0}$ |

## 6.6. Harvard Realization of Some Fibonacci Devices

### 6.6.1. The device for the reduction of the Fibonacci representations to the MINIMAL FORM

The *convolution* and *devolution* devices play an important role in the technical realization of the arithmetical operations over the Fibonacci representations. They can be designed on the basis of the binary register, having special logical circuits to perform the *convolutions* and the *devolutions*. Each digit of the register contains the binary flip-flop (trigger) and logical elements. The operations of *convolution* ($011 \to 100$) and *devolution* ($100 \to 011$) can be performed by means of the inversing of the "flip-flops" (triggers).

One of the possible variants of the Boolean realization of the *device for the reduction of the Fibonacci code to the MINIMAL FORM* is shown in Fig. 6.1.

The device in Fig. 6.1 consists of five $R$–$S$-triggers and the logical elements $AND$, $OR$, which are used to perform the "convolutions". The "convolution" is performed by using the logical gates $AND_1$–$AND_5$ and the corresponding logical gate $OR$, standing before the $R$- and $S$-inputs of the triggers. The logical gate $AND_1$ performs the *convolution* of the first digit to the second digit. Its two inputs are connected with the direct output of the trigger $T_1$ and the inverse output of the trigger $T_2$. The third input is connected with the synchronization input $C$. The logical gate $AND_1$ analyzes the states $Q_1$ and $Q_2$ of the triggers $T_1$ and $T_2$. If $Q_1 = 1$ and $Q_2 = 0$, this means

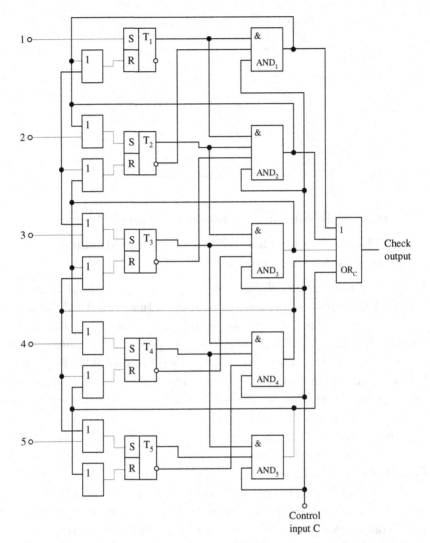

Fig. 6.1. The device for the reduction of the Fibonacci code to the MINIMAL FORM.

that the *convolution* condition is satisfied for the first and second digits. The synchronization signal $C = 1$ causes the appearance of the logical 1 at the output of the gate $AND_1$. The latter causes switching of the triggers $T_1$ and $T_2$. This results in the *convolution* $(01 \rightarrow 10)$.

The logical gate $AND_k$ of the $k$th digit ($k = 2, 3, 4, 5$) performs the *convolution* of the $(k-1)$th and $k$th digits to the $(k+1)$th digit. Its three inputs are connected with the direct outputs of the triggers $T_{k-1}$ and $T_k$ and the inverse output of the trigger $T_{k+1}$. The fourth input is connected with the synchronization input $C$. The logical gate $AND_k$ analyzes the states of $Q_{k-1}$, $Q_k$, and $Q_{k+1}$ of the triggers $T_{k-1}$, $T_k$, and $T_{k+1}$. If $Q_{k-1} = 1$, $Q_k = 1$, and $Q_{k+1} = 0$, this means that the *convolution* condition is satisfied. The synchronization signal $C = 1$ results in switching of triggers $T_{k-1}$, $T_k$, and $T_{k+1}$. The *convolution* of the corresponding digits (011→100) is over.

Note that all the gates $AND_1$–$AND_5$ are connected through the common logical gate $OR_c$ with the check output of the *convolution* register.

The device for the *reduction of the Fibonacci code to the* MINIMAL FORM *in* Fig. 6.1 operates as follows. The input code information is sent to the informational inputs 1–5 of the *convolution* register and enters the $S$-inputs of the triggers through the corresponding logical gates $OR$. Let the initial slate of the convolution register be in the following state:

$$5\ 4\ 3\ 2\ 1$$
$$0\ 1\ 0\ 1\ 1.$$

It is clear that the *convolution* condition is satisfied only for the first, second and third digits. The first synchronization signal $C = 1$ results in the passage of the *convolution* register to the following state:

$$5\ 4\ 3\ 2\ 1$$
$$0\ 1\ 1\ 0\ 0.$$

Here, the *convolution* condition is satisfied for the third, fourth and fifth digits. The next synchronization signal $C = 1$ results in the passage of the *convolution* register to the following state:

$$5\ 4\ 3\ 2\ 1$$
$$1\ 0\ 0\ 0\ 0.$$

The *convolution* is over.

### 6.6.2. The convolution register as a self-checking device

The outputs of the logical gates $AND_1$–$AND_5$ of the *convolution* registers in Fig. 6.1 are connected with the register check output through the common gate $OR$. This output plays the important role of the check output of the *convolution* register.

It follows from the functioning principle of the *convolution* register that the logical 1 appears on the check output only for two situations:

(1) The binary code word, inscribed into the *convolution* register, is not MINIMAL FORM. This means that the *convolution* condition is satisfied at least for one triple of the adjacent triggers of the *convolution* register. This causes the appearance of the logical 1 at the output of the corresponding gate $AND$. Hence, in this case, the appearance of the logical 1 at the check output of the *convolution* register indicates the fact that the *convolution* process is not over. This means that we have a possibility to indicate the termination of the *convolution* process by means of observing the check output of the *convolution* register.

(2) The appearance of the constant logical 1 at the check output is the indication of the fault in the *convolution* register. Hence, the **convolution** register is a natural self-checking device, which is important for the improvement of informational reliability of the *Fibonacci processor*.

There are other variants of implementation of such device described in Ref. [6].

### 6.6.3. Device for checking the MINIMAL FORM

Figure 6.2 shows a device for checking MINIMAL FORM for the case $p = 1$ [6]. The device consists of the $n$ logical gates $AND$. Their outputs are connected with the inputs of the common logic gate $OR$. If the initial Fibonacci representation has a violation of the MINIMAL FORM, that is, has the two adjacent bits of 1 or the bit 1 in the lower digit, logical 1 appears at least at one input of the logical

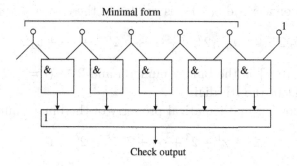

Fig. 6.2. The logical circuit for checking the MINIMAL FORM.

gate *AND*. It results in the appearance of the logical 1 at the output of the common gate *OR* and this logical 1 is the error indication.

## 6.7. Fibonacci Counter for the MINIMAL FORM and Zeckendorf's Theorem

### 6.7.1. The simplest mathematical identities for the Fibonacci numbers

We will use here the following simplest mathematical properties of the classic Fibonacci numbers described in Chapter 1 of Volume I:

$$F_1 + F_2 + \cdots + F_n = F_{n+2} - 1, \tag{6.45}$$

$$F_1 + F_3 + F_5 + \cdots + F_{2k-1} = F_{2k}, \tag{6.46}$$

$$F_2 + F_4 + F_6 + \cdots + F_{2k} = F_{2k+1} - 1. \tag{6.47}$$

### 6.7.2. The Fibonacci code for the case $p = 1$

As well known, the *Fibonacci code* is the following positional representation of natural numbers:

$$N = a_n F_n + a_{n-1} F_{n-1} + \cdots + a_i F_i + \cdots + a_1 F_1, \tag{6.48}$$

where $a_i \in \{0, 1\}$ is the binary numeral and $F_i$ $(i = 1, 2, 3, \ldots, n)$ is the weight of the $i$th digit.

The Fibonacci code (6.48) is similar to the *classical binary code*:

$$N = a_n 2^{n-1} + a_{n-1} 2^{n-2} + \cdots + a_i 2^{i-1} + \cdots + a_1 2^0, \qquad (6.49)$$

where $a_i \in \{0, 1\}$ is the binary numeral and $2^{i-1}$ $(i = 1, 2, 3, \ldots, n)$ is the weight of the $i$th digit.

The following mathematical property of the binary numbers

$$2^0 + 2^1 + 2^3 + \cdots + 2^{n-1} = 2^n - 1 \qquad (6.50)$$

is very important for the designing of the *binary counters*. According to (6.50), the transition of the binary combination $\underbrace{011 \cdots 11}_{n}$ to the next binary combination $\underbrace{100 \cdots 00}_{n}$ is performed with the carryover of the bit 1 from the preceding digit to the next digit. The problem transfer of the bits from the minor to senior digits is the major problem at the designing of the classical binary counters.

By comparing the property (6.50) for the binary numbers to similar properties (6.45)–(6.47) for the *Fibonacci numbers*, we see that the *Fibonacci counter* has much greater opportunities for the creation of the effective algorithms for the Fibonacci counters. This idea underlies the original *Fibonacci counter* presented in this section.

### 6.7.3. Fibonacci representation

The brief notation of the $n$-digit *Fibonacci code* (6.48) consists of the $n$ bits, starting from the senior digit $a_n$, and represents the positive integer $N$ in the following compact form:

$$N = a_n a_{n-1} \cdots a_i \cdots a_2 a_1. \qquad (6.51)$$

The brief notation (6.51) is called the *Fibonacci representation* of the positive integer $N$.

We emphasize that the Fibonacci code (6.48) is similar to the binary code (6.49) because the *Fibonacci representation* (6.51) consists only of the bits 0 and 1. This means that the *Fibonacci counter* and the *Fibonacci micro-processors* are similar to the classical binary counters and micro-processors because they use the same binary elements and binary logic, which is important for micro-electronic technology.

### 6.7.4. Zeckendorf's theorem

Édouard Zeckendorf was born in Liège in 1901. In 1925, Zeckendorf graduated as the medical doctor from the University of Liège and joined the Belgian Army medical center. He was a Belgian medical doctor, an army officer and was enthusiastic about mathematics. In mathematics, he was best known for his study of the *Fibonacci numbers*, in particular, he proved *Zeckendorf's theorem* (Fig. 6.3).

In order to prove Zeckendorf's theorem, let's accept the following theorem without proof.

**Theorem 6.4.** *The arbitrary positive integer $N$ has one and only one representation in the form:*

$$N = F_i + r, \tag{6.52}$$

*where $F_i$ $(i = 2, 3, 4, \ldots)$ is some Fibonacci number and $r$ is some non-negative integer, which satisfies the conditions:*

$$0 \leq r < F_{i-1}. \tag{6.53}$$

Theorem 6.4 can be used for the proof of the following theorem called *Zeckendorf's theorem*.

Fig. 6.3. Édouard Zeckendorf (1901–1983).

**Zeckendorf's theorem.** *Every positive integer can be represented as the unique sum of the non-adjacent Fibonacci numbers.*

The examples of *Zeckendorf's theorem* are the following representations of natural numbers from 38 to 42 by using the *non-adjacent Fibonacci numbers*:

$$38 = 34 + 3 + 1; \quad 39 = 34 + 5; \quad 40 = 34 + 5 + 1;$$

$$41 = 34 + 5 + 2; \quad 42 = 34 + 8. \tag{6.54}$$

*Zeckendorf's theorem* states that such a representation is unique for every natural number.

*Zeckendorf's theorem* is described in Ref. [9] and in the numerous articles published in *The Fibonacci Quarterly*.

## 6.8.    Algorithm of the Fibonacci Counter for the MINIMAL FORM

### 6.8.1.    An example of the Fibonacci counting for the MINIMAL FORM

Table 6.9 demonstrates the example of the *"Fibonacci counting"* for the case of the 8-bit MINIMAL FORM:

$$N = a_8 a_7 a_6 a_5 a_4 a_3 a_2 a_1. \tag{6.55}$$

The analysis of Table 6.9 allows the formulation of the following general peculiarities of the *"Fibonacci counting"*, used only in MINIMAL FORMS:

(1) First of all, we do one important remark. The minor bit $a_1$ of every MINIMAL FORM is identically equal to 0, that is, the MINIMAL FORM (6.55) is written as follows:

$$N = a_8 a_7 a_6 a_5 a_4 a_3 a_2 0. \tag{6.56}$$

This property is valid for the arbitrary $n$-bit MINIMAL FORM, that is, in the general case, we have

$$N = a_n a_{n-1} \cdots a_i a_{i-1} \cdots a_3 a_2 0. \tag{6.57}$$

Table 6.9. Sequential change of the MINIMAL FORMS.

| $n$ | 8 | 7 | 6 | 5 | 4 | 3 | 2 | 1 | $n$ | 8 | 7 | 6 | 5 | 4 | 3 | 2 | 1 |
|---|---|---|---|---|---|---|---|---|---|---|---|---|---|---|---|---|---|
| $N/F_n$ | 21 | 13 | 8 | 5 | 3 | 2 | 1 | 1 | $N/F_n$ | 21 | 13 | 8 | 5 | 3 | 2 | 1 | 1 |
| 0 | 0 | 0 | 0 | 0 | 0 | 0 | 0 | 0 | 13 | 0 | 1 | 0 | 0 | 0 | 0 | 0 | 0 |
| 1 | 0 | 0 | 0 | 0 | 0 | 0 | 1 | 0 | 14 | 0 | 1 | 0 | 0 | 0 | 0 | 1 | 0 |
| 2 | 0 | 0 | 0 | 0 | 0 | 1 | 0 | 0 | 15 | 0 | 1 | 0 | 0 | 0 | 1 | 0 | 0 |
| 3 | 0 | 0 | 0 | 0 | 1 | 0 | 0 | 0 | 16 | 0 | 1 | 0 | 0 | 1 | 0 | 0 | 0 |
| 4 | 0 | 0 | 0 | 0 | 1 | 0 | 1 | 0 | 17 | 0 | 1 | 0 | 0 | 1 | 0 | 1 | 0 |
| 5 | 0 | 0 | 0 | 1 | 0 | 0 | 0 | 0 | 18 | 0 | 1 | 0 | 1 | 0 | 0 | 0 | 0 |
| 6 | 0 | 0 | 0 | 1 | 0 | 0 | 1 | 0 | 19 | 0 | 1 | 0 | 1 | 0 | 0 | 1 | 0 |
| 7 | 0 | 0 | 0 | 1 | 0 | 1 | 0 | 0 | 20 | 0 | 1 | 0 | 1 | 0 | 1 | 0 | 0 |
| 8 | 0 | 0 | 1 | 0 | 0 | 0 | 0 | 0 | 21 | 1 | 0 | 0 | 0 | 0 | 0 | 0 | 0 |
| 9 | 0 | 0 | 1 | 0 | 0 | 0 | 1 | 0 | 22 | 1 | 0 | 0 | 0 | 0 | 0 | 1 | 0 |
| 10 | 0 | 0 | 1 | 0 | 0 | 1 | 0 | 0 | 23 | 1 | 0 | 0 | 0 | 0 | 1 | 0 | 0 |
| 11 | 0 | 0 | 1 | 0 | 1 | 0 | 0 | 0 | 24 | 1 | 0 | 0 | 0 | 1 | 0 | 0 | 0 |
| 12 | 0 | 0 | 1 | 0 | 1 | 0 | 1 | 0 | 25 | 1 | 0 | 0 | 0 | 1 | 0 | 1 | 0 |

In this case, we can exclude from the consideration the minor bit $a_1 = 0$ from the Fibonacci code (6.48) and consider the "truncated" MINIMAL FORMS of the *Fibonacci code* with the digit weights $F_2 = 1, F_3 = 2, F_4 = 3, F_5 = 5, F_6 = 8, \ldots, F_n$, that is,

$$N = a_n F_n + a_{n-1} F_{n-1} + \cdots + a_i F_i + \cdots + a_3 F_3 + a_2 F_2. \quad (6.58)$$

(2) For the arbitrary MINIMAL FORM, the following characteristic property is valid: after each bit of 1 *from the left to the right*, the bit of 0 follows, i.e., the two bits of 1 do not occur near. This property is the reflection of *Zeckendorf's theorem* in the *Fibonacci representations* (6.56) and (6.57). Taking into consideration this fundamental property, Table 6.9 can be represented, as shown in Table 6.10.

## 6.9. General Operational Rules of the Fibonacci Counter for the MINIMAL FORM

The analysis of Table 6.12 allows the establishment of the following general operational rules of the *Fibonacci counting for the MINIMAL*

Table 6.10. The "truncated" MINIMAL FORMS of the Fibonacci 1-code.

| $n$ | 8 | 7 | 6 | 5 | 4 | 3 | 2 | $n$ | 8 | 7 | 6 | 5 | 4 | 3 | 2 |
|---|---|---|---|---|---|---|---|---|---|---|---|---|---|---|---|
| $N/F_n$ | 21 | 13 | 8 | 5 | 3 | 2 | 1 | $N/F_n$ | 21 | 13 | 8 | 5 | 3 | 2 | 1 |
| 0 | 0 | 0 | 0 | 0 | 0 | 0 | 0 | 13 | 0 | 1 | 0 | 0 | 0 | 0 | 0 |
| 1 | 0 | 0 | 0 | 0 | 0 | 0 | 1 | 14 | 0 | 1 | 0 | 0 | 0 | 0 | 1 |
| 2 | 0 | 0 | 0 | 0 | 0 | 1 | 0 | 15 | 0 | 1 | 0 | 0 | 0 | 1 | 0 |
| 3 | 0 | 0 | 0 | 0 | 1 | 0 | 0 | 16 | 0 | 1 | 0 | 0 | 1 | 0 | 0 |
| 4 | 0 | 0 | 0 | 0 | 1 | 0 | 1 | 17 | 0 | 1 | 0 | 0 | 1 | 0 | 1 |
| 5 | 0 | 0 | 0 | 1 | 0 | 0 | 0 | 18 | 0 | 1 | 0 | 1 | 0 | 0 | 0 |
| 6 | 0 | 0 | 0 | 1 | 0 | 0 | 1 | 19 | 0 | 1 | 0 | 1 | 0 | 0 | 1 |
| 7 | 0 | 0 | 0 | 1 | 0 | 1 | 0 | 20 | 0 | 1 | 0 | 1 | 0 | 1 | 0 |
| 8 | 0 | 0 | 1 | 0 | 0 | 0 | 0 | 21 | 1 | 0 | 0 | 0 | 0 | 0 | 0 |
| 9 | 0 | 0 | 1 | 0 | 0 | 0 | 1 | 22 | 1 | 0 | 0 | 0 | 0 | 0 | 1 |
| 10 | 0 | 0 | 1 | 0 | 0 | 1 | 0 | 23 | 1 | 0 | 0 | 0 | 0 | 1 | 0 |
| 11 | 0 | 0 | 1 | 0 | 1 | 0 | 0 | 24 | 1 | 0 | 0 | 0 | 1 | 0 | 0 |
| 12 | 0 | 0 | 1 | 0 | 1 | 0 | 1 | 25 | 1 | 0 | 0 | 0 | 1 | 0 | 1 |

*FORM.* As it follows from Table 6.10, the transition from the *Fibonacci representation* (6.57) of the number $N$ to the *Fibonacci representation* of the next number $N+1$ is done by writing the bit of 1 into the corresponding digit of the next *Fibonacci representation* of the number $N+1$. At the same time, we must adhere to the following rules.

**Rule 1:** The next bit of 1 is written directly into the minor digit of the "truncated" Fibonacci representation (6.57) (that is, to the digit with the index 2) only for the condition, when the bits of the first two digits (with the numbers 2 and 3) in the preceding *Fibonacci representation* of the number $N$ are equal to 0 (the cases, corresponding to the *Fibonacci representations* of the integers 0, 5, 8, 11, 13, 16, 19, 21, 24 in Table 6.10).

**Rule 2:** Let the *Fibonacci representation* of the integer $N$ be in the following form:

| $n$ | $n-1$ | $\cdots$ | $2k+3$ | $2k+2$ | $2k+1$ | $2k$ | $2k-1$ | $2k-2$ | $2k-3$ | $\cdots$ | 4 | 3 | 2 |
|---|---|---|---|---|---|---|---|---|---|---|---|---|---|
| $a_n$ | $a_{n-1}$ | $\cdots$ | $a_{2k+3}$ | 0 | 0 | 1 | 0 | 1 | 0 | $\cdots$ | 1 | 0 | 1 |

We can see that in this *Fibonacci representation*, we have a specific group of the digits, starting from the digit with index 2 and ending with digit with index $(2k+2)$. This group of the digits consists of two parts:

(1) *The group of digits from the digit with index 2 to the digit with index $(2k)$*: Here, the digit with the index 2 contains the bit of 1, but the other digits starting from the digit 3 and ending with the digit with the index $(2k)$ contain the alternating bits 0 and 1 from the right to the left; herewith the bits 1 are in all the digits with the *even* indexes: $2, 4, 6, \ldots, 2k - 2, 2k$.

(2) The group of two digits with the indexes $(2k + 1)$ and $(2k + 2)$ contain the bits 0 in both the digits (see in bold): Note that the remaining group of digits with the indexes from $(2k + 3)$ to $n$ satisfies the condition of the MINIMAL FORM (two bits of 1 do not occur near).

Then, the change of the *Fibonacci representation* of number $N$ on the next *Fibonacci representation* of the number $N + 1$ (when the next bit 1 enters the input of the counter) is realized according to **Rule 2:** *The bit of 1 is recorded into the digit with the index $(2k+1)$, and all the remaining digits to the right of the digit with the index $(2k + 1)$ are nulled;* herewith, the *Fibonacci representation* of the number $N + 1$ takes the following form:

| $n$ | $n-1$ | $\cdots$ | $2k+3$ | $2k+2$ | $2k+1$ | $2k$ | $2k-1$ | $2k-2$ | $2k-3$ | $\cdots$ | 4 | 3 | 2 |
|---|---|---|---|---|---|---|---|---|---|---|---|---|---|
| $a_n$ | $a_{n-1}$ | $\cdots$ | $a_{k+3}$ | **0** | **1** | 0 | 0 | 0 | 0 | $\cdots$ | 0 | 0 | 0 |

Let's prove that this *Fibonacci representation* corresponds to the integer $N + 1$. This follows directly from the formula (6.57) $F_2 + F_4 + F_6 + \cdots + F_{2k} = F_{2k+1} - 1$, that is, the new *Fibonacci representation*, in fact, is the *Fibonacci representation* (in the MINIMAL FORM) of the integer $N + 1$.

**Rule 3:** Let the *Fibonacci representation* of the integer $N$ be in the following form:

| $n$ | $n-1$ | $\cdots$ | $2k+2$ | $2k+1$ | $2k$ | $2k-1$ | $2k-2$ | $2k-3$ | $2k-4$ | $\cdots$ | 4 | 3 | 2 |
|---|---|---|---|---|---|---|---|---|---|---|---|---|---|
| $a_n$ | $a_{n-1}$ | $\cdots$ | $a_{2k+2}$ | **0** | **0** | 1 | 0 | 1 | 0 | $\cdots$ | 0 | 1 | 0 |

We can see that in this *Fibonacci representation*, we have a specific group of digits, starting from the digit with index 3 and ending with digit with the index $(2k+1)$. This group of digits consists of two parts:

(1) *The group of the digits starting from the digit with index* 2 *and ending with the digit with index* $(2k-1)$. Here, the digit with the index 2 is equal to 0, but the other digits represent the alternating bits of 1 and 0; herewith the bits 1 are in all the digits with the *odd* indexes: $3, 5, 7, \ldots, 2k-3, 2k-1$.

(2) *The group of the two digits with the indexes* $2k$ *and* $(2k+1)$, herewith the numerals 0 are in both the digits with the indexes $2k$ and $(2k+1)$ (see in bold). Note that the remaining group of digits with the indexes from $(2k+2)$ to $n$ satisfies the condition of the MINIMAL FORM (the two bits of 1 do not occur near).

Then, the change of the *Fibonacci representation* of number $N$ on the next *Fibonacci representation* of number $N+1$ (when the next bit 1 enters the input of the counter) is realized according to **Rule 3**: *The bit of 1 is recorded into the digit with index 2k, but all the remaining digits to the right are nulled*; herewith, the new *Fibonacci representation* of the number $N+1$ has the following form:

| $n$ | $n-1$ | $\cdots$ | $2k+2$ | $2k+1$ | $2k$ | $2k-1$ | $2k-2$ | $2k-3$ | $2k-4$ | $\cdots$ | 4 | 3 | 2 |
|-----|-------|----------|--------|--------|------|--------|--------|--------|--------|----------|---|---|---|
| $a_n$ | $a_{n-1}$ | $\cdots$ | $a_{k+2}$ | 0 | 1 | 0 | 0 | 0 | 0 | $\cdots$ | 0 | 0 | 0 |

Let's now prove that this *Fibonacci representation* corresponds to the integer $N+1$. It is clear that the weight of the digit with the index $2k$ is equal to $F_{2k}$. On the other hand, according to (6.47), $F_2 + F_4 + F_6 + \cdots + F_{2k} = F_{2k+1} - 1$, the sum of all the Fibonacci numbers with the *odd* indexes $1, 3, 5, \ldots, 2k-1$ is equal to $F_{2k}$. Let's recall that we consider the "*truncated*" *Fibonacci code* (6.58), where the digit with index 1 is absent. Taking this fact into consideration, we can rewrite the identity (6.58) as follows:

$$F_3 + F_5 + \cdots + F_{2k-1} = F_{2k} - F_1 = F_{2k} - 1. \qquad (6.59)$$

Thus, the numerical equivalent of the *nulled* digits is equal to $F_{2k} - 1$, whereas the numerical equivalent of the *nulled* indexes equal $F_{2k}$. It follows from this consideration that the new *Fibonacci representation*, in fact, is the MINIMAL FORM of number $N+1$.

A new operational algorithm of the Fibonacci counter shows that the transition from the *Fibonacci representation* of number $N$ into the *Fibonacci representation* of the next number $N+1$ is carried out per one clock cycle, which includes the record of the bit of 1 into the appropriate digit, and *nulling* the groups of the digits. *This creates prerequisites for the creation of the high-speed Fibonacci counter.* But most importantly, this counter uses only the MINIMAL FORMS for the representation of numbers; these forms are the *"allowed"* *Fibonacci representations* and the main "checking" forms of the *Fibonacci code*.

In order to ease the description of the structural circuit of the *Fibonacci counter*, we are changing renumbering of the digits in the *"truncated"* *Fibonacci code* (6.58) and will use the *Fibonacci code*, in which the numeration of digits begins with the digit with index 1:

$$N = a_n F_{n+1} + a_{n-1} F_n + \cdots + a_{i-1} F_i + \cdots + a_2 F_3 + a_1 F_2.$$

$$(6.60)$$

We will use this numeration in Section 6.10 in the Fibonacci counter for the MINIMAL FORMS.

## 6.10. Boolean Realization of the Fibonacci Counter for the MINIMAL FORM

### 6.10.1. A general description

The analysis of the above *Fibonacci counter algorithm* shows that we should use five operational blocks for the realization of the *Fibonacci counter*: *Register, Disposition Block, Block for Analysis, Block for Error Checking, and Block for Zero Installation*. These blocks allow the designing of the *high-speed and noise-immune Fibonacci counter*. Its noise immunity is achieved due to the presence of the *"forbidden"* *states* of the *counter*; its high speed is achieved as a result of the absence of the previous carryovers, needed in the classical binary counters, and also the micro-operations of the *convolutions* and *devolutions*, used in the known *Fibonacci counters* [16]. These properties give some benefits compared with classical binary counters and the known *Fibonacci counters* [16].

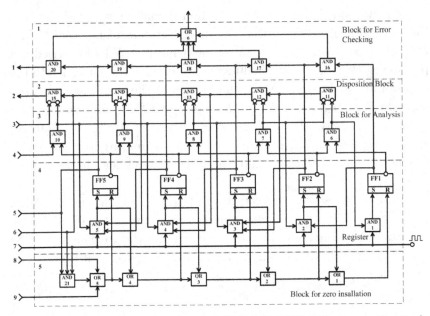

Fig. 6.4. The Fibonacci counter for the MINIMAL FORM (Boolean realization).

Let's consider the operation of the *Fibonacci counter* on the example of the 5-bit *Fibonacci counter* (Fig. 6.4).

The *Fibonacci counter* in Fig. 6.4 consists of the *Register* (the 5 RS-flip-flops FF1–FF5 with the logical circuits AND1–AND5), *Block for Analysis* (the logical circuits AND6–AND10), *Disposition Block* (the logical circuits AND11–AND15), *Block for Error Checking* (the logical circuits AND16–AND20 and OR6), *Block for Zero Installation* (the logical circuits AND21 and OR1–OR5). All these blocks have the regular structure and can be of unlimited length. Accordingly, the *Fibonacci counters* can have an unlimited length. To increase the length of the *Fibonacci counter*, for example, on the one digit (the sixth digit), it is only necessary to link corresponding outputs of the sixth digit to the inputs 1–9 by analogy as it was done previously in Fig. 6.4 at the connection of the fourth digit with the fifth digit. Herewith, we should put the logical circuit AND21 to the input of the sixth digit flip-flop. By analogy, we can design the seventh, eighth digits and so on up to the *n*th digit of the *Fibonacci counter*.

## 6.10.2. The Register

The *Register* (the 5 RS-flip-flops FF1–FF5 with the logical circuits AND1–AND5) is needed to memorize the counter states. For the counter in Fig. 6.4, we use the RS-flip-flops FF1–FF5 as the most simple, although we may also use of the other kinds of flip-flops, which perform this function, for example, the JK-flip-flops. The outputs of the RS-flip-flops FF1–FF5 are linked to the inputs of the logical circuits AND6–AND10 of the *Block for Analysis*, but the inputs of the *RS*-flip-flops FF1–FF5 are linked with the *Block for Zero Installation* and the bus clock.

## 6.10.3. Block for Analysis, Block for Error Checking, Disposition Block, and Block for Zero Installation

The *Block for Analysis* controls the outputs of the pairs of the neighboring flip-flops of the *Register*, starting from FF1 and FF2, which can be in the three "allowed" states 00, 01 or 10. Let's recall that the state 11 for the two adjacent flip-flops is "forbidden" in accordance with the algorithm of the *Fibonacci counter*. In order to identify the "forbidden" states, we introduced into the *Fibonacci counter* the *Block for Error Checking*, which signal about the error state of the *Fibonacci counter*.

To analyze the location of the bits 1's in the *Fibonacci counter*, we introduced the *Disposition Block*. The *Disposition Block* together with the *Block for Analysis* control that the flip-flops of the *Register* go to the next correct state of the counter.

The *Block for Zero Installation* sets all the flip-flops of the *Register* into the state 0 for two cases. The first case, when any flip-flop of the *Register* is switched over into the state 1; herewith, all the flip-flops on the left should be switched over into the state 0. For the second case, the *Block for Zero Installation* sets all the flip-flops of the *Register* into the state 0 after the end of the counting cycle.

The structure in Fig. 6.4 is quite simple and it is easy to trace that the transition from the MINIMAL FORM of the number $N$ to the MINIMAL FORM of the number $N + 1$ is performed strictly in accordance to Table 6.10.

## 6.11. Functioning of the Fibonacci Counter for the MINIMAL FORM

### 6.11.1. The initial state

The *Fibonacci counter* in Fig. 6.4 operates as follows. First, the *Fibonacci counter* is in the *initial state*, when all the flip-flops FF1–FF5 of the *Register* are installed into the *zero state*. For this case, the logical signals of 0 from the direct outputs of the flip-flops FF1–FF4, according to Fig. 6.4, enter the inputs of the logical circuits AND2–AND5, respectively. Note that the inverse outputs of the flip-flops FF1–FF2 are connected with the inputs of the logical circuit AND6, and the inverse outputs of the flip-flops FF2–FF3 are connected with the inputs of the logical circuit AND7 and so on up to the logical circuits AND10, as shown in Fig. 6.4; this means that on the outputs of the logical circuits AND6–AND10 are formed the signals of 1. The logical signals of 1 from the outputs of the logical circuits AND6–AND10 enter to the corresponding inputs of the logical circuits AND1–AND5, respectively.

**The first step:** According to Fig. 6.4, the first clock pulse enters the inputs of the logical circuits AND1–AND5. However, if the *Fibonacci counter* is in the *zero state* 00000, only the flip-flop FF1 through the logical circuit AND1 can be switched over into the state of 1 because all the remaining logical circuits AND2–AND5 are blocked by the 0-signals entered from the direct outputs of the flip-flops FF2–FF5. As a result, the *Fibonacci counter* is installed into the state 00001, corresponding to integer 1.

**The second step:** The transition of the flip-flop FF1 into state 1 leads to the following situation. The signal 0 appears on the output of the logical circuit AND6 and then on the input of the logical circuit AND1. This signal forbids the passage of the next clock pulse to the $S$-input of the flip-flop FF1. The logical circuit AND2, on the contrary, is open for the passing of the next clock pulse to the $S$-input of the flip-flop FF2. This means that the next (second) clock pulse switches over the flip-flop FF2 into the state of 1, while the second clock pulse, through the logical circuit OR1, switches over the flip-flop FF1 to the state 0. As a result, we get a new state of the *Fibonacci counter* 00010 corresponding to integer 2.

**The third step:** Let's consider the situation with the logical circuit AND3 after the *second step*. This logical circuit has four inputs. Since the *Fibonacci counter* is in the state 00010, then the signal of 1 from the direct output of the flip-flop FF2 enters the first input of the logical circuit AND3. Since the flip-flops FF3 and FF4 are in the state 0, then the signal of 1 from the output of the logical circuit AND8 enters the second input of the logical circuit AND3. Finally, let's consider the situation with the logical circuits AND6, AND7, and AND11. Since the *Fibonacci counter* is in the state 00010, then the input signals of the logical circuits AND6 and AND7 are equal to 0 and the input signal of the logical circuit AND11 is equal to 1. Thus, the signal 1 enters the third input of the logical circuit AND3. This means that the third clock pulse, through the logical circuit AND3, enters the $S$-input of the flip-flop FF3 and switches over it into the state of 1. Besides, this logical 1, through the logical circuits OR2 and then OR1, enters the $R$-inputs of the flip-flops FF2 and FF1 and switches over them into the state of 0. As a result, we get the new state of the *Fibonacci counter* 00100 corresponding to the integer 3.

**The fourth step:** When the Fibonacci counter is in the state 00100, the logical signals of 1 appear on the inverse outputs of the flip-flops FF2 and FF1. These logical signals lead to the appearance of the signal of 1 on the output of the logical circuit AND6. On the other hand, the signal of 1 from the output of the logical circuit AND6 enters the input of the logical circuit AND11 of the *Disposition Block*. This leads to the appearance of the logical 0 at the output of the logical circuit AND11. The signal of 0 from the output of the logical circuit AND11 leads to the appearance of the signal of 0 on the outputs of the logical circuits AND12, AND13, and AND14. These signals of 0 enter the inputs of the logical circuits AND3, AND4, and AND5. Finally, the signal of 0 on the inverse output of the flip-flop FF3, which is in the state of 1, causes the logical signal of 0 on the output of the logical circuit AND7 and on the input of the logical circuit AND2. This means that the next (fourth) clock pulse does not change the states of the flip-flops FF2–FF5. However, the flip-flop FF1 is switched over into the state of 1 with the fourth clock signal. As a result, the counter turns into the state 00101 that corresponds to integer 4.

**The fifth step:** Let's now consider, how the transition is performed from number 4 = 00101 to the next number 5 = 01000. For the situation 00101, the flip-flops FF1 and FF3 are in the state of 1. This means that the signals of 0 appear on the outputs of the logical circuits AND6–AND8. These signals enter the inputs of the logical circuits AND1–AND3. The signals of 0 on the outputs of the logical circuits AND7 and AND8 lead to the appearance of the signal of 1 on the output of the logical circuit AND12, and then this signal enters the input of the logical circuit AND4. In addition, the signal of 1 enters another input of the logical circuit AND4 from the direct output of the flip-flop FF3, which is in the state of 1. Also, the signal of 1 enters the third input of the logical circuit AND4 from the output of the logical circuit AND9. As a result, the next (fifth) clock pulse sets the flip-flop FF4 in the state of 1, and then, through the logical circuits OR3–OR1, sets the flip-flops FF3, FF2, and FF1 into the state of 0. As a result, the *Fibonacci counter* turns into the state 01000 corresponding to number 5 = 01000.

**The sixth step:** Since the flip-flops FF1, FF2, and FF3 are in the state 0, then the signals of 0 enter the corresponding inputs of the logical circuits AND2–AND4 from the direct outputs of these flip-flops. Besides, the signals of 1 enter the inputs of the logical circuits AND6 and AND7 from the inverse outputs of the flip-flops FF1, FF2, and FF3, which are in the state of 0. The signals of 1 enter the inverse inputs of the logical circuit AND11 and they lead to the appearance of the signal of 0 on the output of the logical circuit AND11. This signal of 0 leads to the appearance of the signal of 0 on the output of the logical circuit AND12 and then AND13. The signal of 0 enters the input of the logical circuit AND5 from the output of the logical circuit AND13. Since the flip-flops FF1 and FF2 are in the state of 1, the signal of 1 appears on the output of the logical circuit AND6. This signal enters the input of the logical circuit AND1. Thus, only the flip-flop FF1 can switch over with the clock pulse. The next (sixth) clock signal switches over the flip-flop FF1 into the state of 1 and does not change the states of the flip-flops FF2–FF5. As a result, the *Fibonacci counter* turns into the state 01001 corresponding to number 6 = 01001.

**The seventh step:** Further, by analogy, the next (seventh) clock pulse turns the *Fibonacci counter* into the state 01010 corresponding to number 7 = 01010.

**The eighth step:** For this situation, there appears the signal of 1 on the inputs of the logical circuit AND5 and the next (eights) clock pulse turns the *Fibonacci counter* into the state 10000 corresponding to number 8 = 10000.

**The last steps:** This process will continue up to the transition of the *Fibonacci counter* into the state 10101 corresponding to the maximal number 12 = 10101, which can be represented with the 5-bit Fibonacci code in the MINIMAL FORM. For this case, the two signals of 1 enter the inputs of the logical circuit AND21 from the direct input of the flip-flop FF5 and the logical circuit AND14 of the *Disposition Block*. The next clock pulse, through the logical circuits AND21 and OR5–OR1, turns the flip-flops FF1–FF5 into the initial state 00000. After that, the counter is ready for the new counting cycle.

## 6.11.2. Increasing the Fibonacci counter capacity

As we mentioned above, the additional inputs and outputs of counters 1–9 are intended for an unlimited increase of the Fibonacci counter capacity *on the left* and *on the right*. For this purpose, we can use the standard set of logical circuits for the one digit of the counter. This standard structure consists of the one two-input logical circuit OR of the *Block for zero installation*, the one flip-flop and the one logical circuit AND of the *Register*, the one two-input logical circuit AND of the *Block for Analysis*, the one two-input logical circuit AND of the *Disposition Block*, and the one 2-digit logical circuit of the *Block for Error Checking*.

### 6.11.3. Functioning of the Block for Error Checking

The *Block for Error Checking* is, in fact, the external block regarding the *Fibonacci counter* and does not affect the counting algorithm. Its task is to identify errors in the Fibonacci counter. The random appearance of the two adjacent bits of 1 in the flip-flops of the *Register* leads to the fact that the "error signal" of 1 appears on one of the logical circuits AND of the *Block for Error Checking*. This "error signal" passes through the logical circuit OR6 to the output of the *Block for Error Checking*. For this case, the *Fibonacci counter* can start the new counting cycle or, if there is a duplicate counter, can be switched over to the duplicate counter.

## 6.12. The Main Peculiarities of the Fibonacci Counter for the MINIMAL FORM

### 6.12.1. A range of number representation for the MINIMAL FORM

Let's consider the $n$-digit MINIMAL FORMS (6.51) of the Fibonacci code (6.48). It is clear that we can represent all the integer numbers from the minimal number $N_{\min} = 0 = \underbrace{00 \cdots 0}_{n}$ to the maximal number $N_{\max}$.

In order to calculate the maximal number $N_{\max}$, which can be represented by the $n$-digit MINIMAL FORMS of the Fibonacci code (6.48), we carefully study Table 6.9.

Let's consider the adjacent Fibonacci representations of the numbers taken from Table 6.9: $4 = 1010$ and $5 = 10000$, $7 = 10100$ and $8 = 100000$, $12 = 101010$ and $13 = 1000000$ (see in bold in Table 6.9), and so on. Note that the numbers 5, 8 and 13 are the *Fibonacci numbers* $F_5 = 5$, $F_6 = 8$, $F_7 = 13$. It is easy to see that the number $4 = 1010$ is the maximal number, which can be represented by the four-digit MINIMAL FORM, number $7 = 10100$ is the maximal number, which can be represented by the 5-bit MINIMAL

FORM, and number $12 = 101010$ is the maximal number, which can be represented by the 6-bit MINIMAL FORM. All of them are calculated according to the formulas: $4 = 5 - 1$, $7 = 8 - 1$, $12 = 13 - 1$.

These arguments lead us to the following theorem, which can be proved by the induction on $n$.

**Theorem 6.5.** *The maximal number $N_{\max}$, which can be represented by using $n$-digit MINIMAL FORMS of the Fibonacci code* (6.48), *is equal to $F_{n+1} - 1$; in this case, the range of number representation from $N_{\min} = 0 = \underbrace{00 \cdots 0}_{n}$ to $N_{\max} = F_{n+1} - 1$ is equal to $F_{n+1}$, where $F_{n+1}$ is the Fibonacci number.*

## 6.12.2. A redundancy of the Fibonacci code for the MINIMAL FORM

Let $n = m + k$ be a number of digits of the redundant code, where $m$ is the number of the *data bits* and $k$ is the number of the *redundant bits*. Then, the *relative redundancy* of the code can be calculated according to the following formula [16, 17]:

$$R = \frac{k}{n} = \frac{n - m}{n} = 1 - \frac{m}{n}. \qquad (6.61)$$

We now calculate the *relative redundancy* of the Fibonacci code (6.48) for the MINIMAL FORM by using the formula (6.61). According to Theorem 6.5, by using the $n$-digit MINIMAL FORMS of the Fibonacci code (6.48), we can represent $F_{n+1}$ integers in the range from $N_{\min} = 0 = \underbrace{00 \cdots 0}_{n}$ to $N_{\max} = F_{n+1} - 1$. It is clear that we need $m \approx \log_2 F_{n+1}$ bits to represent this number range in the classical binary code (6.49). Substituting this expression into (6.61), we get the following formula for the calculation of the *relative redundancy* of the Fibonacci code (6.48) for the MINIMAL FORM:

$$R = 1 - \frac{m}{n} = 1 - \frac{\log_2 F_{n+1}}{n}. \qquad (6.62)$$

To simplify the formula (6.62), we can use the so-called *Binet's formula* for the Fibonacci numbers [11]:

$$F_{n+1} = \frac{\left(\frac{1+\sqrt{5}}{2}\right)^{n+1} - \left(\frac{1-\sqrt{5}}{2}\right)^{n+1}}{\sqrt{5}} \approx \frac{\left(\frac{1+\sqrt{5}}{2}\right)^{n+1}}{\sqrt{5}} = \frac{\Phi^{n+1}}{\sqrt{5}}, \quad (6.63)$$

where $\Phi = \frac{1+\sqrt{5}}{2} \approx 1.618$ is the golden ratio [11].

By using (6.63), we can represent the formulas (6.62) as follows:

$$R = 1 - \frac{n\log_2 \Phi + \log_2 \Phi - \frac{1}{2}\log_2 5}{n}. \quad (6.64)$$

With increasing $n$, the expression (6.64) converges to the following:

$$R = 1 - \log_2 \Phi \approx 0.306. \quad (6.65)$$

It is important to emphasize that the *relative code redundancy* (6.65) for the Fibonacci code (6.48) does not depend on the length of the Fibonacci representation and in the limit is equal to 0.306 (30.6%).

### 6.12.3. Error-detecting ability of the Fibonacci code

We draw the following considerations, which are of general character and applicable for arbitrary redundant code [16]. If we use the MINIMAL FORMS for the number representation, then the number of "allowed" $n$-digit Fibonacci representations $\{a_n a_{n-1} \cdots a_i a_{i-1} \cdots a_2 a_1\}$ is equal to the number of all the $n$-digit MINIMAL FORMS $F_{n+1}$, while the number of all possible $n$-digit binary representations $\{a_n a_{n-1} \cdots a_i a_{i-1} \cdots a_2 a_1\}$ is equal to $2^n$, then the number of the "forbidden" $n$-digit Fibonacci representations is equal to the difference $2^n - F_{n+1}$.

The errors are detected in all cases, when the "allowed" Fibonacci representations (MINIMAL FORMS) pass into the "forbidden" Fibonacci representation. If the "allowed" Fibonacci representation passes into the other "allowed" Fibonacci representation, this is a case of "undetectable error."

Table 6.11. Error-detecting ability of
the Fibonacci code for the MINIMAL
FORM.

| $n$ | $2^n$ | $F_{n+1}$ | $S_d\%$ |
|-----|-------|-----------|---------|
| 10 | 1 024 | 89 | 91.31 |
| 16 | 65 536 | 1 597 | 97.56 |
| 24 | 16 777 216 | 75 025 | 99.55 |

A number of all possible transitions is equal to $F_{n+1} \times 2^n$; here, the number of "detectable" transitions is equal to $F_{n+1} \times (2^n - F_{n+1})$. The ratio

$$S_d = \frac{F_{n+1} \times (2^n - F_{n+1})}{F_{n+1} \times 2^n} = 1 - \frac{F_{n+1}}{2^n} \qquad (6.66)$$

characterizes the *error-detecting ability* of the Fibonacci code (6.48).

The numerical values of the coefficient $S_d$, given by (6.66), for the various values of $n$, are presented in Table 6.11.

By comparing the *classical error-correcting codes* with the *Fibonacci code* (6.48), we should not forget that the *Fibonacci code* (6.48) is, first of all, the positional numeral system. By using the *Fibonacci code* (6.48), we can design the *noise-immune Fibonacci computers* and *micro-processors*, which allows the performing of all the data transformations and provides for the checking of these transformations. If we use the *Fibonacci code* (6.48) in computers, we don't need to convert the non-redundant code (the binary code) into the redundant code and conversely. This automatically solves encoding–decoding problem and leads to the simplification of computational structures.

Thus, the *Fibonacci code* (6.48) exceeds the *classical error-correcting codes*, first of all, in qualitative relation (because the classical error-correcting codes are not numeral systems). On the other hand, the *Fibonacci code* (6.48) is the *redundant binary numeral system*, which allows the detection of errors in computers. This means that the *Fibonacci code* (6.48) exceeds the *classical binary numeral system* regarding error-detecting ability.

### 6.12.4. The main advantages of the Fibonacci counter for the MINIMAL FORM

The main peculiarity of the Fibonacci counter in Fig. 6.4, compared with the *well-known Fibonacci counters* [16], consists in the fact that *this Fibonacci counter functions only in the MINIMAL FORM of the Fibonacci code* (6.48). All other forms of the number representation in the *Fibonacci code* (6.48) are "forbidden", and their occurrence denotes the signal of errors, which may occur in the *Fibonacci counter* under the influence of various internal and external effects. As can be seen from Table 6.11, the coefficient of error detection increases with the increase of the *Fibonacci counter capacity*. For example, the 24-bit *Fibonacci counter* has a coefficient of error detection equal to 99.55%. Recall that the potential error-detecting coefficient for the code with parity check, used widely in computers today, is equal only to 50%. On the other hand, the device for error detection for the Fibonacci code (see Fig. 6.2) is simpler than a similar device for the code with a parity check.

The Fibonacci counter in Fig. 6.4 has a fairly high speed. This is achieved due to the fact that the carryovers from digit to digit are absent. This improves the speed in comparison with the known binary error-correcting counters and the known Fibonacci counters [16] based on the "convolutions" and "devolutions". A time delay in this device only occurs in the chain of the logic circuits AND in the *Disposition Block* and the *Block for Zero Installation* (see Fig. 6.4). However, this delay occurs within one clock cycle in parallel with switching over one flip-flop, but not during several cycles, as it usually occurs in the noise-immune binary counters. Regarding the speed, this counter is comparable to the classical binary counter with ripple-through carryover, but it differs from it with high noise immunity.

As follows from Fig. 6.4, the proposed counter has a *high homogeneity* of logical elements and connections between them which is important for micro-electronics and nano-electronics. This is another advantage of the *Fibonacci counter* in Fig. 6.4.

Thus, the proposed counter has *high noise immunity* and *informational reliability* (see Table 6.11), it has a sufficiently *big speed* and *regular structure*. Therefore, it makes sense to recommend this *Fibonacci counter* for use in the digital devices, which require high *performance, noise immunity,* and *reliability,* such as devices for the *frequency measurement* and the *control systems for the mission-critical applications.*

### 6.12.5. Significance of Zeckendorf's theorem for future development of computing and micro-electronic technology

As mentioned above, in 1939, the Belgian medical doctor, army officer and amateur of mathematics Eduardo Zeckendorf (1901–1983) had proved the interesting theorem, known in the *Fibonacci numbers theory* as *Zeckendorf's theorem* [9]. This theorem is a mathematical result of great importance for computer technology. This theorem concerns the representation of positive integers as the sum of Fibonacci numbers. Thus, Eduardo Zeckendorf predicted in his famous theorem a principally new way in the development of computer technology — *Fibonacci computers*, which began to develop in modern science starting since the 1960s.

The Fibonacci counter in Fig. 6.4 is only the first step for designing the reliable Fibonacci computers, based on *Zeckendorf's theorem* and on the conception of the MINIMAL FORM, based on this theorem.

The next step in the development of the Fibonacci computers is designing all the operational devices of the *Fibonacci computers,* in particular, ALU, electronic memory and microchips, on the basis of the MINIMAL FORM. In the *Fibonacci computers,* ALU, the memory and the microchips, all the input and output data should be represented in the MINIMAL FORM, which provides the continuous checking of errors on all the stages of data transformations in the *Fibonacci computers* (arithmetical operations, storage and data transmission).

### 6.13. The Main Principles of the Fibonacci Computer

We can formulate the following main principles of the *Fibonacci computer*:

(1) In general, we can use in the *Fibonacci computers* the *redundant Fibonacci p-codes* (6.1). However, the Fibonacci 1-code, having the *least code redundancy*, is the most suitable in terms of hardware cost and implementation.

(2) We use the MINIMAL FORM of the *Fibonacci 1-code* for error detection at all the stages of data transformation, including arithmetic operations, data transfer, and storage.

The *Fibonacci computer* has a number of important advantages in comparison with classic binary computers. The most important of these is the *high ability to detect errors in the functional units of the Fibonacci computer.*

### 6.13.1. "Soft" and "hard" errors

As well known, all errors, arising in functional devices of computer, can be divided into two groups: (1) *soft errors*, which result from the *random* effects on the electronic elements and *hard errors*, which result from the *constant* failures of electronic elements. Both types of errors are dangerous and may lead to *false data* on the computer output.

As for the *hard errors*, they can be detected by the register for the reduction of the Fibonacci code to the MINIMAL FORM. This register is an important device of all arithmetical units and thanks to this device, all the *Fibonacci arithmetical devices* have become *self-checking devices*. This is the first important advantage of the *Fibonacci computer.*

### 6.13.2. Formula for error-detecting ability of the soft errors

Let us now consider the error detection in an important computer unit such as the *electronic memory*. The *error-detecting ability* of the *Fibonacci p-codes* is determined by the relationship between the *allowed* and *forbidden* code combinations. Note that the set of *allowed* combinations of the $n$-digit *Fibonacci p-code* coincides with the set of $n$-digit MINIMAL FORMS. This means that the number of $n$-bit *allowed* combinations is equal to $N_1 = F_p(n+1)$. Then, the number of *forbidden* code combinations is equal to $N_2 = 2^n - F_p(n+1)$, where $2^n$ is the number of possible $n$-digit binary combinations. Then the *error-detecting ability* $S_d$ of the Fibonacci $p$-code (for the case of the MINIMAL FORM) is determined as follows:

$$S_d(p) = \frac{N_2}{2^n} = 1 - \frac{F_p(n+1)}{2^n}. \qquad (6.67)$$

For example, for the case of the 24-digit *Fibonacci p-codes* ($p = 1$ and $p = 2$), we have the following numerical values for the range of representations, respectively:

$$\boldsymbol{p = 1}: \quad F_1(25) = 62215, \quad 2^{24} = 16777216, \qquad (6.68)$$

$$\boldsymbol{p = 2}: \quad F_1(25) = 6450, \quad 2^{24} = 16777216. \qquad (6.69)$$

By using (6.67), (6.68), and (6.69), we can calculate the *error-detecting ability* of the 24-digit *Fibonacci 1- and 2-codes*, respectively, as follows:

$$\begin{aligned} S_d(p = 1) &= 0.9963 \ (99.63\%), \\ S_d(p = 2) &= 0.9996 \ (99.96\%). \end{aligned} \qquad (6.70)$$

### 6.13.3. Fibonacci Parity Code

As seen from the example (6.70), the *potential error-detecting ability* of the Fibonacci $p$-codes is high enough. In order to improve the potential error-detecting ability of the Fibonacci code, we can use the so-called *Fibonacci Parity Code* (FPC) by adding the parity bit $a_{par}$ to the MINIMAL FORM of the *Fibonacci code*:

$$\underbrace{a_n a_{n-1} \cdots a_i \cdots a_2 a_1}_{MF} \underbrace{a_{par}}_{PB}. \qquad (6.71)$$

The Fibonacci parity code (FPC) (6.71) significantly improves the error-detecting ability of the Fibonacci $p$-code. In this case, the main feature of FPC is ensuring the 100% detection of all the *odd-digit* errors, in particular, the single-digit errors. It is easy to prove that the potential error-detecting ability of FPC (6.71) is calculated by the following formula:

$$S_d(FPC) = 1 - \frac{F_p(n+1)}{2^{n+1}}. \tag{6.72}$$

For example, for the case of the 24-digit Fibonacci $p$-codes ($p = 1$ and $p = 2$), we have the following data ranges for the FPC (for $p = 1$ and $p = 2$), respectively:

$$F_1(25) = 62215, \quad 2^{25} = 33554432, \tag{6.73}$$

$$F_2(25) = 6450, \quad 2^{25} = 33554432. \tag{6.74}$$

By using (6.72) for the cases (6.73) and (6.74), we can calculate the error-detecting ability of the FPC (for $p = 1$ and $p = 2$), respectively, as follows:

$$S_d(FPC, p = 1) = 0.9981459 \ (99.8\%), \tag{6.75}$$

$$S_d(FPC, p = 2) = 0.9998078 \ (99.98\%). \tag{6.76}$$

This means that the FPC can provide the continuous detection of the errors in Fibonacci microcontroller or micro-processor at the various stages of storage, transmission and processing of data with the *error-detecting coefficient* equal to 99.8–99.98%.

### 6.13.4. Energy consumption and power dissipation in ROM

It is known that for certain types of electronic memory (EM), there is some *asymmetry* between the bits 1 and 0 at their storage for different kinds of electronic memory, for example, ROM. In particular, the recording of the bit of 1 and its reading from the ROM requires more energy consumption than for the bit of 0. From this point of view, the MINIMAL FORM of the *Fibonacci representations* is the optimal binary representations from the point of view of the *energy consumption* and the *power dissipation* because the bits of 1 are separated always with bits of 0 (in general, in the MINIMAL FORM, the two bits of 1 are separated by no less than $p$ bits of 0).

It is clear, in the array of the MINIMAL FORMS, the bits of 1 and 0 are distributed non-uniformly; in this case, the number of bits of 0 always exceeds the number of bits of 1. This creates "comfortable" conditions for the electronic memory, in particular, for the ROM, from the point of view of *energy consumption* and *power dissipation*. The ROMs with the fusible links or electrically programmable ROMs are the examples of such kind of electronic memory. For such ROMs only the bits of 1 determine the *energy consumption* and the *power dissipation*.

In order to estimate the decrease of the *energy consumption* in the ROM, when the data are stored in the MINIMAL FORM of the *Fibonacci p-codes*, we consider the following reasonings. When we store in the ROM the numbers, represented with the $m$-digit binary code, the *maximal energy consumption* appears at the recording and reading of the following binary code combination:

$$N_{\max}(m) = \underbrace{11\cdots1}_{m}. \tag{6.77}$$

If the storage of the one bit of 1 demands the *energy consumption* $P_1$, then we can express the *maximal energy consumption* $P_{\max}(m)$ for the storage of the code combination (6.77) as follows:

$$P_{\max}(m) = mP_1. \tag{6.78}$$

For the $m$-digit binary code, we can represent $2^m$ *positive integers*, that is, the range of the *number representation* is equal to

$$D = 2^m. \tag{6.79}$$

If we represent the positive integers in the MINIMAL FORM of the *Fibonacci p-code* (6.1), then, due to its code redundancy, for the storage of the same range of the positive integers, given by (6.79), we need to increase the number of digits of the Fibonacci $p$-code (6.1) proportionally to its *relative code redundancy* $r$. In this case, the number of Fibonacci's digits $n$ for the storage of the *number range* (6.79) is approximately equal to

$$n = m(1 + r). \tag{6.80}$$

The formula (6.80) shows that the number of digits $n$ in the *Fibonacci p-code* (6.1) increases $(1 + r)$ times in comparison with the number of digits of $m$ for the binary system necessary for the representation of the same number range (6.79). In this case, the $n$-digit *Fibonacci representation* of the maximal number $N_{max}$ in the MINIMAL FORM of the *Fibonacci p-code* (6.1) is written as follows:

$$N_{max} = 1\underbrace{00\ldots0}_{p}1\underbrace{00\ldots0}_{p}\ldots1\underbrace{00\ldots0}_{p}\ldots1\underbrace{00\ldots0}_{p}. \qquad (6.81)$$

In particular, for the case $p = 1$, we have

$$N_{max} = 1010\ldots10\ldots10. \qquad (6.82)$$

We can see that the *Fibonacci representation* (6.81) consists of the $k$ groups of the kind $1\underbrace{00\ldots0}_{p}$, which contains one bit of 1 and the $p$ bits of 0, where $k$ is the number of bits of 1 in the *Fibonacci representation* (6.81). By using (6.80) and (6.82), we can express the number of $k$ as follows:

$$k = \frac{n}{p+1} = \frac{m(1+R)}{p+1}. \qquad (6.83)$$

It follows from (6.83) that the *maximal energy consumption* for the storage of the $n$-digit Fibonacci representation is determined by the following expression:

$$P^f_{max} = kP_1 = \frac{m(1+r)}{p+1}P_1. \qquad (6.84)$$

Let's now consider the ratio:

$$\beta = \frac{P_{max}}{P^f_{max}} = \frac{p+1}{1+R}. \qquad (6.85)$$

Since $R < 1$ and $p \geq 1$, then the coefficient $\beta$ describes the decreasing *energy consumption* in the ROM if we use the *Fibonacci p-codes* (6.1).

Table 6.12 sets forth the values of $\beta$ for the cases $p = 1, 2$.

Table 6.12. The values of $\beta$.

| $p$ | 1 | 2 |
|-----|---|---|
| $R$ | $\approx 0.33(33\%)$ | $\approx 0.5(50\%)$ |
| $\beta$ | 1.5 | 2 |

Thus, for the case $p = 2$ the improvement in *energy consumption* is about 2 times, despite the fact that the number of digits in the *Fibonacci 2-code* increases by about 1.5 times compared with the classical binary system. It follows from Table 6.12 that the gain in *energy consumption* in the ROM increases with the increase in $p$. This result could have great significance for the future of the nanocomputers and micro-electronics, where the decrease in the energy consumption and the optimizing power dissipation is becoming one of the central problems.

## 6.14. Fibonacci Arithmetic Based on the Basic Micro-Operations

### 6.14.1. The basic micro-operations

As we mentioned above, the main distinction of the *Fibonacci 1-code* (6.6) compared to the *binary code* (6.5) is the *multiplicity* of the *Fibonacci representations* of the one and the same positive integer. By using the above micro-operations of the *devolution* (011 → 100) and the *convolution* (100 → 011), we can change the forms of the *Fibonacci representations* of the one and the same positive integer. This means that the bits of 1 in the *Fibonacci representation* (6.6) of the one and the same number can move to the left or to the right along the Fibonacci representation (6.6) of the same number by using the micro-operations of the *devolution* (011 → 100) and the *convolution* (100 → 011). Recall once more that the fulfilment of these micro-operations does not change the number itself, that is, we will get the different Fibonacci *representations* of the one and the same number. This fact allows development of the original approach to the *Fibonacci arithmetic* based on the so-called *basic micro-operations*.

Let's introduce the following four *basic micro-operations* used to fulfill logical and arithmetical operations over the binary words:

| Convolution | Devolution | Replacement | Absorption |
|---|---|---|---|
| $100 \leftarrow 011$ | $100 \rightarrow 011$ | $\begin{bmatrix} 1 & 0 \\ \downarrow & = \\ 0 & 1 \end{bmatrix}$ | $\begin{bmatrix} 1 & 0 \\ \updownarrow & = \\ 1 & 0 \end{bmatrix}$ |

$$(6.86)$$

Note that the noise-immune *Fibonacci arithmetic*, based on the above micro-operations (6.86), is described for the first time in the brochure [23].

Note that the *convolutions* and *devolutions*, shown in the table (6.86), are the simple code transformations, which are performed over the adjacent three bits of the *Fibonacci representation* of the one and the same number $N$ in the *Fibonacci 1-code* (6.6).

In addition to the micro-operations of the "*convolution*" and the "*devolution*", discussed above, we consider two new micro-operations: "*replacement*" and "*absorption*", presented in the table (6.86).

The micro-operation of the *replacement* $\begin{bmatrix} 1 & 0 \\ \downarrow & = \\ 0 & 1 \end{bmatrix}$ is the two-placed micro-operation, which is fulfilled over the same digits of the two registers: the *top* register $A$ and the *lower* register $B$. Let's now consider the case, when the register $A$ has the bit of 1 in the $k$th digit and the register $B$ has the bit of 0 in the same $k$th digit (the condition for the *replacement*). The micro-operation of the *replacement* consists in the moving of the bit of 1 from the $k$th digit of the *top* register $A$ to the $k$th digit of the lower register $B$. Note that this operation can be fulfilled only for the condition if the bits of the $k$th digits of the registers $A$ and $B$ are equal to 1 and 0, respectively.

The micro-operation of the *absorption* $\begin{bmatrix} 1 & 0 \\ \updownarrow & = \\ 1 & 0 \end{bmatrix}$ is the two-placed micro-operation for the condition, when the bits of 1 are in the $k$th digits of the *top* register $A$ and the lower register $B$. This micro-operation consists in the mutual annihilation of the bits of 1 in the *top* and lower registers $A$ and $B$. After the fulfillment of the micro-operation of the *absorption*, the bit of 1 is replaced with the bit of 0.

It is necessary to pay attention to the following "technical" peculiarity of the above "*basic micro-operations*". At the register interpretation of these micro-operations, each micro-operation may be fulfilled by means of the *inversion* of flip-flops, involved in the micro-operation. This means that each micro-operation reduces to flip-flops' switching.

### 6.14.2. Logical operations

We can demonstrate the possibility of fulfilling the simplest logical operations by means of the above *basic micro-operations* (6.86). Let's now perform all possible *replacements* from the *top* register $A$ to the *lower* register $B$:

$$A \;=\; 1\ 0\ 0\ 0\ 1\ 0\ 1$$
$$\downarrow \qquad\quad \downarrow$$
$$B \;=\; 0\ 1\ 0\ 1\ 0\ 0\ 1$$
$$\overline{\phantom{B \;=\; 0\ 1\ 0\ 1\ 0\ 0\ 1}}$$
$$A' \;=\; 0\ 0\ 0\ 0\ 0\ 0\ 1$$
$$B' \;=\; 1\ 1\ 0\ 1\ 1\ 0\ 1$$

As a result of the *replacement*, we get the two new binary words $A'$ and $B'$. We can see that the binary word $A'$ is the logical *conjunction* ($\wedge$) of the initial binary words $A$ and $B$, that is,

$$A' = A \wedge B,$$

and the binary word $B'$ is the logical *disjunction* ($\vee$) of the initial binary words $A$ and $B$, that is,

$$B' = A \vee B.$$

The logical operation of the *module 2 addition* is fulfilled by means of the simultaneous fulfillment of all the possible *replacements* and *absorptions*:

$$A \ = \ 1 \ 0 \ 1 \ 0 \ 0 \ 1 \ 1 \ 0 \ 1$$
$$\updownarrow \quad \downarrow \quad \ \ \downarrow \ \downarrow \quad \ \updownarrow$$
$$B \ = \ 1 \ 1 \ 1 \ 0 \ 0 \ 1 \ 1 \ 0 \ 1$$

$$A' \ = \ 0 \ 0 \ 0 \ 0 \ 0 \ 0 \ 0 \ 0 \ 0 \ = \ \text{const } 0$$
$$B' \ = \ 0 \ 1 \ 1 \ 0 \ 0 \ 1 \ 1 \ 0 \ 0 \ = \ A \oplus B$$

We can see that the results of this code transformation are the two new binary words $A' = \text{const } 0$ and $B' = A \oplus B$. It is clear that the binary word $A' = \text{const } 0$ plays the role of the *checking* of the logical operation of *addition by module* 2, which is important for the computer applications.

The logical operation of the *code A inversion* reduces to the fulfillment of the *absorptions* over the initial binary word $A$ and the special binary word $B = \text{const } 1$:

$$A \ = \ 1 \ 0 \ 1 \ 0 \ 0 \ 1 \ 1 \ 0 \ 1$$
$$\updownarrow \quad \updownarrow \quad \ \ \updownarrow \ \updownarrow \quad \ \updownarrow$$
$$B \ = \ 1 \ 1 \ 1 \ 1 \ 1 \ 1 \ 1 \ 1 \ 1 \ = \ \text{const } 1$$

$$A' \ = \ 0 \ 0 \ 0 \ 0 \ 0 \ 0 \ 0 \ 0 \ 0 \ = \ \text{const } 0$$
$$B' \ = \ 0 \ 1 \ 0 \ 1 \ 1 \ 0 \ 0 \ 1 \ 0 \ = \ \ \ \overline{A}$$

The binary word $A' = \text{const } 0$ plays a role of the *checking* of the logical operation of the *inversion*, which is important for the computer applications.

### 6.15. Fibonacci Summation and Subtraction by Using "Basic Micro-Operations"

#### 6.15.1. Fibonacci summation

The idea of summation of the two numbers $A$ and $B$ by using the *basic micro-operations* consists of the following. We have to move all the binary 1's from the *top* register $A$ to the *lower* register $B$. With this purpose, we use the micro-operations of the *replacement*: the *devolution* and the *convolution*. The result is formed in the register $B$.

For example, let's summarize the Fibonacci representations $A_0 = 010100100$ and $B_0 = 001010100$ as follows.

**The first step** of the *Fibonacci summation* consists in the *replacement* of all the possible bits of 1 from the top register $A$ to the lower register $B$:

$$
\begin{array}{rccccccccc}
A_0 = & 0 & 1 & 0 & 1 & 0 & 0 & 1 & 0 & 0 \\
& & \downarrow & & \downarrow & & & & & \\
B_0 = & 0 & 0 & 1 & 0 & 1 & 0 & 1 & 0 & 0 \\
\hline
A_1 = & 0 & 0 & 0 & 0 & 0 & 0 & 1 & 0 & 0 \\
B_1 = & 0 & 1 & 1 & 1 & 1 & 0 & 1 & 0 & 0
\end{array}
\qquad (6.87)
$$

For this, we apply the micro-operations of the *replacement* to all the digits of the initial *Fibonacci representations* $A_0 = 010100100$ and $B_0 = 001010100$. However, this can be fulfilled only for those digits, where the condition of the *replacement* $\begin{bmatrix} 1 & 0 \\ \downarrow = \\ 0 & 1 \end{bmatrix}$ is satisfied.

**The second step** is the fulfillment of all the possible *devolutions* in the *Fibonacci representation* $A_1$ and all the possible *convolutions* in the *Fibonacci representation* $B_1$, that is,

$$
\begin{array}{rccccccccc}
A_1 = & 0 & 0 & 0 & 0 & 0 & 0 & 1 & 0 & 0 \\
B_1 = & 0 & 1 & 1 & 1 & 1 & 0 & 1 & 0 & 0 \\
& & & & \Downarrow & & & & & \\
A_2 = & 0 & 0 & 0 & 0 & 0 & 0 & 0 & 1 & 1 \\
B_2 = & 1 & 0 & 0 & 1 & 1 & 0 & 1 & 0 & 0
\end{array}
\qquad (6.88)
$$

**The third step** is the *replacement* $\begin{bmatrix} 1 & 0 \\ \downarrow = \\ 0 & 1 \end{bmatrix}$ over all the possible bits of 1 from the top register $A$ to the lower register $B$:

$$
\begin{array}{rccccccccc}
A_2 = & 0 & 0 & 0 & 0 & 0 & 0 & 0 & 1 & 1 \\
& & & & & & & & \downarrow & \downarrow \\
B_2 = & 1 & 0 & 0 & 1 & 1 & 0 & 1 & 0 & 0 \\
\hline
A_3 = & 0 & 0 & 0 & 0 & 0 & 0 & 0 & 0 & 0 \\
B_3 = & 1 & 0 & 0 & 1 & 1 & 0 & 1 & 1 & 1
\end{array}
\qquad (6.89)
$$

The summation is over because all the bits of 1 have moved from the top register $A$ to the lower register $B$. After the reduction of the *Fibonacci representation* $B_3$ to the MINIMAL FORM, we get the sum $B_3 = A_0 + B_0$ represented in the MINIMAL FORM:

$$B_3 = 10\,\boxed{011}\,\boxed{011}\,1 = 1010010\,\boxed{01} = 101001010 = A_0 + B_0.$$

Thus, the *Fibonacci summation* reduces to the sequential fulfillment of the micro-operations of the *replacement* over the two initial *Fibonacci representations* $A_0$ and $B_0$, the micro-operations of the *convolution* over the *intermediate Fibonacci representation* $B$ and the micro-operations of the *devolution* over the *intermediate Fibonacci representation* $A$. The process of the *Fibonacci summation* continues until the moment, when the *Fibonacci representation* in the *top* register $A$ becomes equal to *zero*.

### 6.15.2. Fibonacci subtraction

The idea of the *Fibonacci subtraction* of the two *Fibonacci representations*, kept in the registers $A$ and $B$, consists of the following. We use the three basic micro-operations, the *absorption*, the *devolution*, and the *convolution*, to transfer all the bits of 1 from one register to another. The *Fibonacci subtraction* ends in the moment, when in one of the registers, the *zero* code combination $\underbrace{000\cdots0}_{n}$ will be formed.

The sign of the result of the subtraction depends on the fact, in which one of the registers $(A$ or $B)$ prove to be the result of the subtraction. If the result of the subtraction proved to be in the top register $A$, then the sign of the subtraction is a "positive number." Otherwise, the result of the subtraction is a "negative number."

### 6.15.3. The example of the Fibonacci subtraction

We need to get the result of the subtraction of the two *Fibonacci representations* kept in the *top* and *lower* registers $A$ and $B$:

$$D = A_0 - B_0 = 1010010000 - 101010010,$$

where $D$ means the *difference* or the *remainder*.

The *Fibonacci subtraction* over the two *Fibonacci representations* $A_0$ and $B_0$, kept in the top and lower registers $A$ and $B$, is realized as follows.

**The first step** of the *Fibonacci subtraction* involves the *absorptions* over all the possible bits of 1 in the *initial Fibonacci representations* $A_0$ and $B_0$:

$$A_0 = 1\ 0\ 1\ 0\ 0\ 1\ 0\ 0\ 0$$

$$\updownarrow \quad \updownarrow$$

$$B_0 = 1\ 0\ 1\ 0\ 1\ 0\ 0\ 1\ 0 \tag{6.90}$$

$$\overline{\phantom{B_0 = 1\ 0\ 1\ 0\ 1\ 0\ 0\ 1\ 0}}$$

$$A_1 = 0\ 0\ 0\ 0\ 0\ 1\ 0\ 0\ 0$$

$$B_1 = 0\ 0\ 0\ 0\ 1\ 0\ 0\ 1\ 0$$

**The second step** involves the *devolutions* over the intermediate *Fibonacci representations* $A_1$ and $B_1$:

$$A_1 = 0\ 0\ 0\ 0\ 0\ 1\ 0\ 0\ 0$$

$$B_1 = 0\ 0\ 0\ 0\ 1\ 0\ 0\ 1\ 0$$

$$\Downarrow \tag{6.91}$$

$$A_2 = 0\ 0\ 0\ 0\ 0\ 0\ 1\ 1\ 0$$

$$B_2 = 0\ 0\ 0\ 0\ 0\ 1\ 1\ 0\ 1$$

**The third step** involves the *absorptions* over the *Fibonacci representations* $A_2$ and $B_2$:

$$A_2 = 0\ 0\ 0\ 0\ 0\ 0\ 1\ 1\ 0$$

$$\updownarrow$$

$$B_2 = 0\ 0\ 0\ 0\ 0\ 1\ 1\ 0\ 1$$

$$\overline{\phantom{B_2 = 0\ 0\ 0\ 0\ 0\ 1\ 1\ 0\ 1}}$$

$$A_3 = 0\ 0\ 0\ 0\ 0\ 0\ 0\ 1\ 0$$

$$B_3 = 0\ 0\ 0\ 0\ 0\ 1\ 0\ 0\ 1$$

**The fourth step** involves the *devolutions* over the *Fibonacci representations* $A_3$ and $B_3$:

$$A_3 = 0\,0\,0\,0\,0\,0\,0\,\boxed{1\,0}$$
$$B_3 = 0\,0\,0\,0\,0\,\boxed{1\,0\,0}\,1$$
$$\Downarrow \qquad\qquad (6.92)$$
$$A_4 = 0\,0\,0\,0\,0\,0\,0\,0\,1$$
$$B_4 = 0\,0\,0\,0\,0\,0\,1\,1\,1$$

**The fifth step** involves the *replacements* over the *Fibonacci representations* $A_4$ and $B_4$:

$$A_4 = 0\,0\,0\,0\,0\,0\,0\,0\,1$$
$$\downarrow$$
$$B_4 = 0\,0\,0\,0\,0\,0\,1\,1\,1$$
$$\Downarrow \qquad\qquad (6.93)$$
$$A_5 = 0\,0\,0\,0\,0\,0\,0\,0\,0$$
$$B_5 = 0\,0\,0\,0\,0\,0\,1\,1\,0$$

**The sixth step** involves the *convolutions* over the final *Fibonacci representation* $B_5$ and its reduction to the MINIMAL FORM:

$$B_5 = 0\,0\,0\,0\,0\,\boxed{0\,1\,1}\,0 = 000001000. \qquad (6.94)$$

The *Fibonacci subtraction* is completed. Since the result (6.94) of the *Fibonacci subtraction* is kept in the *lower* register $B$, then the result of the *Fibonacci subtraction* is the *negative number*.

## 6.16. A Conception of the Fibonacci Arithmetical Processor for the Noise-Immune Computations

### 6.16.1. Noise-immune computations

In modern computer science, there is a need for fault-tolerant and noise-immune processors. What is the distinction between these two important ways of designing the high-reliable computers and processors? It is well known that the computer program is realized through the processor. The processor consists of flip-flops, which are connected with the combinative logic. In this case, the realization of the program reduces to the flip-flops switching. Unfortunately, it is impossible to eliminate computer errors arising as a result of the faulty functioning of the computer elements. Nevertheless, it is necessary to distinguish two types of errors in computer elements. The first type is the so-called *constant failures*, when the elements *"fall outside the order"* constantly. The second type is the so-called *alternate failures*, when the elements *"fall outside the order"* temporarily, that is, in accidental moments, while in some other moments, the computer elements function correctly. The second type of failures is called the *malfunction*. Processor's *malfunctions* appear under the influence of different internal and external noise factors in the computer elements and their electrical circuits. Thus, fault-tolerant computers are intended for the elimination of the *constant failures* that may appear in the processor and its elements and units during their exploration. The noise-immune processors and computers are intended for the elimination of the *malfunctions* that may appear in the computer elements during their exploration. It is clear that the problem of the designing of the noise-immune computers and processors is a very actual problem of modern computer science. For example, many modern cryptosystems are based on the computations in very large finite fields. The hardware realization of such computational units or processors requires thousands of logical gates. It is very difficult and costly to design the high-reliable processors, which yield error-free results. It means that the problem of the designing of the processors for the noise-immune computations is a very important problem for the reliable cryptosystems and other mission-critical computer systems.

It was proved experimentally that the intensity of *malfunctions* (or the *random failures*) in the computer elements in the switching regime is bigger in two or three exponents of the intensity of the elements, which are in the *stable* states. It follows from this reasoning that the *malfunction of triggers in the switching regime is the most probable reason for the unreliable functioning of the computer processors*. This is why the designing of the self-checking automata, which can allow effective detection of the flip-flops malfunctions in the switching regime, is one of the most actual problems in designing the noise-immune computers and processors for mission-critical applications.

## 6.16.2. Checking the basic micro-operations

The basic idea of the designing of the self-checking Fibonacci processors consists the following. It is necessary to design the effective system of the checking of the basic micro-operations in the process of their functioning. Let's now demonstrate a possibility of the realization of this idea by using the above *basic micro-operations* (*convolution, devolution, replacement, and absorption*), used in the noise-immune Fibonacci arithmetic.

We pay our attention to the following "technical" peculiarity of the above *basic micro-operations*. At the register interpretation of these micro-operations, each micro-operation may be realized by means of the *inversion* of flip-flops, involved in the micro-operation. This means that each micro-operation is realized technically by means of flip-flop switching.

Let's now evaluate the *potential ability* of the *basic micro-operations* to detect errors, which may appear during the micro-operations realization. As well known, the *potential error-detecting ability* is determined by the ratio between the number of detectable errors and the general number of all possible errors. Let's explain the essence of our approach to the detection of errors in the above micro-operations with the example of the micro-operation of the *convolution*:

$$100 \leftarrow 011. \tag{6.95}$$

The *convolution* is fulfilled over the three-digit binary code combination 011 as shown in (6.95). It is clear that there are $2^3 = 8$ possible transitions, which can arise at the fulfillment of the micro-operation (6.95). Note that the only one of them, given by (6.95), is *correct*, that is, it is the *unmistakable* transition. The code combinations,

$$\{011, 100\}, \tag{6.96}$$

which are involved in the *unmistakable* transition (6.95), are called the *allowed* code combinations for the *convolution*. The remaining code combinations, which can appear during the *convolution* (6.95)

$$\{000, 001, 010, 101, 110, 111\}, \tag{6.97}$$

are called the *prohibited* code combinations.

The idea of the error detection consists in the following. If during the fulfillment of the micro-operation (6.95), one of the *prohibited* code combinations (6.97) appears, this fact is the *indication* of error. Note that if the *erroneous transition*

$$011 \rightarrow 011, \tag{6.98}$$

then the *allowed* code combination 011 passes into the same *allowed* code combination 011, which can be interpreted as the case of the *undetectable error*.

Let's now consider the other erroneous situations, which can appear at the fulfillment of the micro-operation (6.95):

$$011 \Rightarrow \left\{\begin{matrix} 0 & 1 & 1 \\ 0 & 0 & 0 \\ 0 & 0 & 1 \\ 0 & 1 & 0 \\ 1 & 0 & 1 \\ 1 & 1 & 0 \\ 1 & 1 & 1 \end{matrix}\right\}. \tag{6.99}$$

Among them, only the *erroneous transition* (6.98) is *undetectable* because the code combination 011 is the *allowed* code combination. All the remaining *erroneous transitions* (6.99) are *detectable*.

Let's analyze the transition (6.98) from the arithmetical point of view. It is clear that the essence of the *erroneous transition* (6.98) consists in the *repetition* of the same code combination 011.

If we analyze this transition from the arithmetical point of view, we can see that this transition does not destroy the numerical information and does not influence the outcome of the arithmetical operations. Hence, the *erroneous transition* (6.98) does not belong to the errors of *catastrophic character*. It can delay maybe only the data processing. All the remaining *erroneous transitions* from (6.99) destroy the numerical information and can hence lead to the *errors of catastrophic character*.

The main conclusion, following from this consideration, consists in the fact that *the set of "catastrophic" code combinations from* (6.97) *coincides with the set of detectable code combinations* taken from (6.99). This means that all the "catastrophic" transitions for the *convolution* are *detectable*. We emphasize once again that the *undetectable transition* (6.98) does not destroy numerical information and, therefore, from the arithmetical point of view cannot belong to the *erroneous transitions* of the *catastrophic character*. This *undetectable* transition is only delaying the data processing.

Thus, it follows from this consideration that we can design by using this idea the computer device for the fulfillment of the *convolution* with the *absolute* (i.e., 100%) potential ability to detect all catastrophic transitions, which may appear at the realization of the *convolution*.

We can have a similar conclusion for other *basic micro-operations*. But the fulfillment of some data processing algorithm in the Fibonacci processor, based on the *basic micro-operations*, reduces to the sequential fulfillment of certain basic micro-operations at every step of the algorithm. *Since the checking of circuits for the realization of the* "basic micro-operations" *has theoretically the "absolute" error-detecting ability, the possibility of designing an arithmetical self-checking Fibonacci processor, which has the "absolute" error-detecting ability (100%) for the "catastrophic" errors, arising in the noise-immune Fibonacci processor at the flip-flops' switching, follows from this consideration.*

### 6.16.3. The hardware realization of the noise-immune Fibonacci processor

The noise-immune Fibonacci processor is based on the *principle of the "cause–effect"* described in Ref. [23] (see Fig. 6.5).

The essence of the *principle of the "cause–effect"* is demonstrated in Fig. 6.5. The initial information (the *"cause"*), which is subjected to data processing, is transformed into the "result" by using some micro-operations. After that, we transform the *"result"* (the *"effect"*) to the initial information (the *"cause"*) and then check that the *"effect"* fits with its *"cause"*. For example, at the fulfillment of the *convolution* for the binary combination 011 (the *"cause"*) (100 ← 011), we get the new binary combination 100 (the *"effect"*), which is a necessary condition for the fulfillment of the *devolution* (100 → 011). This means that the correct fulfillment of the *convolution* leads to the condition for the *devolution*. Analogously, the correct fulfillment of the *devolution* leads to the condition for the *convolution*. It follows from this consideration that the micro-operations of the *convolution* and the *devolution* are mutually checked.

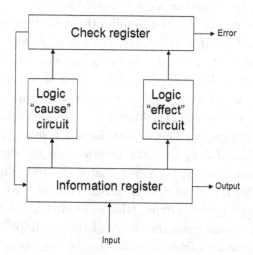

Fig. 6.5. The block diagram of the Fibonacci device for the realization of the principle of the "cause–effect".

These conclusions are true for all the above *basic micro-operations* represented in the table (6.86). For the "register interpretation", the obtaining of the correspondence between the "*cause*" and the "*effect*" is realized by using the "*check register*". The "*cause*" sets up the corresponding "*check flip-flop*" into the state of 1 and the correct fulfillment of the micro-operation (the "*effect*" fits to the "*cause*") overthrows the "check flip-flop" into the state 0. If the "*effect*" does not fit with the "*cause*" (the micro-operation was carried out incorrectly), then the "*check flip-flop*" remains in the state 1, which is the indication of the error.

If we analyze the "*causes*" and the "*effects*" for every *basic micro-operation*, we can determine that every "*effect*" is the *inversion* of its "*cause*", that is, all micro-operations could be realized by means of the *inversion* of flip-flops, involved in the micro-operation.

The block diagram of the *Fibonacci device* for the realization of the *principle of "cause–effect"* is shown in Fig. 6.5. The device in Fig. 6.5 includes the *information* and *check* registers, which are connected by means of the logic "*cause*" and "*effect*" circuits. The code information, entering the *information* register through the "input," is analyzed by the logic "*cause*" circuit.

Let's suppose that we need to fulfill the *convolution* for the binary combination in the *information register*. Let some flip-flops $T_{k-1}$, $T_k$, $T_{k+1}$ of the *information register* be in the state 011, i.e., the condition for the *convolution* is satisfied for this group of flip-flops. Then, the logic "*cause*" circuit (the logic circuit for the *convolution* for this example) results in writing the logic 1 into the corresponding flip-flop $T_k$ of the *check register*. The written logic 1 results in the *inversion* of flip-flops $T_{k-1}$, $T_k$, $T_{k+1}$ of the *information register* by using the back connection, that is, their new states are 100. This means that the condition for the *devolution* is satisfied for this group of flip-flops. Then, the logic "*effect*" circuit (the logic circuit for the *devolution* for this example) analyzes the states of the flip-flops $T_{k-1}$, $T_k$, $T_{k+1}$ of the *information register* and overthrows the same flip-flop $T_k$ of the *check register* to the initial state of 0. Overthrowing the flip-flop $T_k$ of the *check register* into the initial state of 0 confirms that the "*cause*" (011) fits with its "*effect*" (100), that is, the micro-operation of the *convolution* is correct.

Hence, if we get the code word of $00\cdots0$ in the *check register* after the end of all micro-operations, then all *"causes"* fit with their *"effects"*, that is, all the micro-operations are correct. If the *check register* contains at least one logic 1 in some flip-flop, then at least one *basic micro-operation* is not correct. The logic 1's in the flip-flops of the *check register* cause the error signal of 1 at the output *"Error"* of the device in Fig. 6.5. The signal of 1 at the output *"Error"* prohibits the use of the data on the *"Output"* of the *Fibonacci device* in Fig. 6.5.

The most important advantage of the *check principle of the "cause–effect"*, which is realized in the *Fibonacci device* in Fig. 6.5, is the detection of the *error* in the moment of its appearance. The correction of the *error* in the micro-operation is realized by the *repetition* of this micro-operation.

Hence, *the above approach, based on the principle of the "cause–effect", permits one to detect and then to correct data by means of repetition of all "catastrophic" errors arising at the moment of flip-flop switching with* 100% *guaranty.*

A more detailed description of all the benefits of this principle of implementation of the noise-immune *Fibonacci processor* is given in Ref. [23].

## 6.17. Evaluation of the Results Obtained in Chapter 6

(1) The most significant advantage of the *Fibonacci numeral systems* in comparison to other methods of the introduction of code redundancy into computer systems is that they are the generalizations of the classical binary system and retain all the arithmetical and technical advantages of the binary system (*simple rules of the arithmetical operations, simplicity of the number comparison, iterative, or uniform arithmetic structures*).

(2) In the well-known methods of the introduction of the code redundancy into computer systems (e.g., the error-correcting codes), we always assume the presence of at least two codes: the *source code* for representing the numbers and the *codes*

*for the detection and correction of errors*; this creates the so-called *encoding–decoding problem*. The complexity of encoding–decoding, and the need to take into consideration the errors in these devices in many cases can negate the benefits of the error-correcting codes. In the *Fibonacci numeral systems*, the *encoding–decoding problem* does not arise because the *Fibonacci representations* are used simultaneously to *represent numbers*, *to perform the arithmetic operations*, and *to detect the errors*. Besides the error detection with high detection coefficient, *Fibonacci numeral systems* have other useful properties, such as the solution to the *self-synchronization problem* in the serial code sequence. The *self-synchronization property* allows creation of the high-performance serial data transmission systems.

(3) Another additional feature is the improvement of the *energy consumption* and the *power dissipation*. It is shown that if we store the *Fibonacci representations* in the MINIMAL FORM, the gain in the *energy consumption* for the worst case presides the traditional binary code 1.5 and 2 times for the Fibonacci $(p = 1)$- and $(p = 2)$-codes, respectively. Property of MINIMAL FORM (every bit of 1 is surrounded at the left and the right by at least $p$ bits of 0) is useful for improving the *power dissipation*. The solution of this problem is particularly important for certain types of micro-electronic and nano-electronic memory.

(4) The totality of the above advantages of the *Fibonacci numeral systems* (the *high coefficient of error detection*, the *properties of the self-synchronization*, the *improvement of energy consumption*, and *the power dissipation*) gives us a right to single out the *Fibonacci numeral systems* as the unique multi-functional method of *introduction of code redundancy*, which exceeds all the well-known methods, in particular, the traditional *error-correcting codes*, first of all in the qualitative relation.

(5) **Proposed areas of applications**:

   (i) SPACE: spacecraft control and onboard systems, satellites, etc.

   (ii) ENERGY: the nuclear power stations and other energy facilities.

(iii) TRANSPORT: aircraft, trains, subway, and cars.

(iv) AUTOMATION: banks, factories, and public services.

(v) MEDICAL: computerized medical equipment.

(vi) COMMUNICATION: telephone and Internet.

(vii) SCIENCE: research computers and super computers.

(viii) MEDIA: TV and radio broadcasting.

(ix) ROBOTICS: control systems for the robots and robotic tools.

# Chapter 7

# Codes of the Golden $p$-Proportions

## 7.1. Definition of the Codes of the Golden $p$-Proportions and Their Partial Cases

### 7.1.1. The classic binary system and codes of the golden $p$-proportions

The classic binary system has the following interpretation. Let's consider the infinite set of the binary segments with the following length:

$$\{2^n, 2^{n-1}, \ldots, 2^i, \ldots, 2^0, 2^{-1}, \ldots, 2^{-i}, \ldots, 2^{-(n-1)}, 2^{-n}\}, \qquad (7.1)$$

where $i = 0, \pm 1, \pm 2, \pm 3, \ldots$.

By using (7.1), we can represent every real number $A$ as follows:

$$A = \sum_i a_i 2^i, \qquad (7.2)$$

where $a_i$ is the binary numeral of the $i$th digit, $2^i$ is the weight of the $i$th digit, $i = 0, \pm 1, \pm 2, \pm 3, \ldots$.

The digit weights of the *binary system* (7.2) are connected by two binary "arithmetical" properties:

$$2^i = 2 \times 2^{i-1} \quad \text{(the binary \emph{multiplicative} property)}, \qquad (7.3)$$

$$2^i = 2^{i-1} + 2^{i-1} \quad \text{(the binary \emph{additive} property)}, \qquad (7.4)$$

which underlie the *"binary arithmetic"*.

The *binary multiplicative property* (7.3) is the basis of *binary multiplication*, the *left and right shifts* of the binary words, and the *representation of numbers with "floating point"*. The *binary additive property* (7.4) underlies the *binary summation* and the *binary subtraction* (through the use of the *inverse* and the *additional* codes [135]).

The binary code of the real number $A$, which is determined by (7.1), assumes the following generalization. Let's consider the set of the "golden" segments with the following length:

$$\{\Phi_p^n, \Phi_p^{n-1}, \ldots, \Phi_p^0 = 1, \Phi_p^{-1}, \ldots, \Phi_p^{-k}, \ldots\}, \qquad (7.5)$$

where $\Phi_p$ is the *golden p-ratio*, the positive root of the golden $p$-ratio equation

$$x^{p+1} = x^p + 1.$$

By using (7.5), we can get the following positional method of the real number representation:

$$A = \sum_i a_i \Phi_p^i, \qquad (7.6)$$

where $A$ is the *positive real number*, $a_i \in \{0, 1\}$ is the *binary numeral* of the $i$th digit, $\Phi_p^i$ is the *weight* of the $i$th digit, $\Phi_p$ is the *base* of the numeral system (7.6), $p = 0, 1, 2, 3, \ldots$ is the *given integer*, $i = 0, \pm 1, \pm 2, \pm 3, \ldots$.

### 7.1.2. The partial cases of the codes of the golden $p$-proportions

Let's now consider in more detail the above method of the real number representation, which is given by (7.6). First of all, we note that the formula (7.6) sets forth a theoretically infinite number of the binary positional representations of the real numbers because every $p = 0, 1, 2, 3, \ldots$ "generates" its own method of the "golden" representation of the real numbers in the form (7.6).

The *base* of the numeral system is one of the fundamental notions of the *positional numeral system*. The analysis of the sum (7.6) shows that the *base* of the numeral system (7.6) is the *golden p-ratio* $\Phi_p$, the positive root of the golden $p$-ratio equation, $x^{p+1} = x^p + 1$. This is why the representation of the real numbers $A$ in the form (7.6) is called the *codes of the golden p-proportions of the real numbers A* [19].

Let's note the following fact. Except the case of $p = 0$ ($\Phi_{p=0} = 2$), all the remaining golden $p$-proportions $\Phi_p$ are irrational numbers. It follows from this fact that the *codes of the golden p-proportions* (7.6) are the binary $(0,1)$ numeral systems with irrational bases $\Phi_p$ for the cases $p > 0$.

Note that for the case $p = 0$, the *codes of the golden p-proportions* (7.6) reduce to the *classical binary code* (7.2) and for the case $p = 1$, they reduce to *Bergman's system* (8.1). It is clear that Bergman's system (8.1) has the most practical significance because this numeral system with irrational base $\Phi = \frac{1+\sqrt{5}}{2}$ is the simplest for technical implementation.

## 7.2. Conversion of Numbers from Traditional Numeral Systems to Bergman's System

### 7.2.1. The table method

There are two general methods of number conversion from one numeral system to another. The first method is called the *table method* and the second method is called the *analytic method*. The *table method* is based on the preliminary construction in computers of the special table for the *codes of the golden p-proportions*. This method can be realized in computers by means of the special constant electronic memory. In this case, the *codes of the golden p-proportions* of the number $N$ are kept in the memory by the address, which is a *classical binary representation of the numbers N*. Table 7.1 shows the table method of number conversion for the case $p = 1$ (Bergman's system).

Table 7.1. Table method of number conversion into Bergman's system.

| Address $N$ | $\Phi^4$ | $\Phi^3$ | $\Phi^2$ | $\Phi^1$ | $\Phi^0$ | $\Phi^{-1}$ | $\Phi^{-2}$ | $\Phi^{-3}$ | $\Phi^{-4}$ |
|---|---|---|---|---|---|---|---|---|---|
| $0 = 0000$ | 0 | 0 | 0 | 0 | 0. | 0 | 0 | 0 | 0 |
| $1 = 0001$ | 0 | 0 | 0 | 0 | 1. | 0 | 0 | 0 | 0 |
| $2 = 0010$ | 0 | 0 | 0 | 1 | 0. | 0 | 1 | 0 | 0 |
| $3 = 0011$ | 0 | 0 | 1 | 0 | 0. | 0 | 1 | 0 | 0 |
| $4 = 0100$ | 0 | 0 | 1 | 0 | 1. | 0 | 1 | 0 | 0 |
| $5 = 0101$ | 0 | 1 | 0 | 0 | 0. | 1 | 0 | 0 | 1 |
| $6 = 0110$ | 0 | 1 | 0 | 1 | 0. | 0 | 0 | 0 | 1 |
| $7 = 0111$ | 1 | 0 | 0 | 0 | 0. | 0 | 0 | 0 | 1 |
| $8 = 1000$ | 1 | 0 | 0 | 0 | 1. | 0 | 0 | 0 | 1 |
| $9 = 1001$ | 1 | 0 | 0 | 1 | 0. | 0 | 1 | 0 | 1 |
| $10 = 1010$ | 1 | 0 | 1 | 0 | 0. | 0 | 1 | 0 | 1 |
| $11 = 1011$ | 1 | 0 | 1 | 0 | 1. | 0 | 1 | 0 | 1 |

Note that the *top* row of Table 7.1 contains the golden ratio powers of the kind $\Phi^k$ $(k = 0, \pm1, \pm2, \pm3, \ldots)$.

## 7.2.2. Conversion of the fractional numbers

Let's now consider the analytic method of conversion of the fractional numbers to their *"golden" representations* in Bergman's system (8.1). This method is widely used in the classical numeral systems. Its essence consists in the fulfillment of certain arithmetical operations in the initial numeral system for obtaining numerals in the new numeral system.

Let's suppose that the "golden" representation of the fractional number $A$ has the following form:

$$A = a_{-1}\Phi^{-1} + a_{-2}\Phi^{-2} + \cdots + a_{-n}\Phi^{-n} = 0.a_{-1}a_{-2}\cdots a_{-n}.$$

$$(7.7)$$

Let's now suppose that the fractional number $A$ is represented in the MINIMAL FORM. Then, by multiplying the fractional number (7.7) by base $\Phi$, we get the following result:

$$A \times \Phi = a_{-1} + a_{-2}\Phi^{-1} + \cdots + a_{-n}\Phi^{-n+1} = a_{-1}.a_{-2}\cdots a_{-n},$$

$$(7.8)$$

where $a_{-1}$ is the *integer part* of the product $A \times \Phi$ and the sum

$$A_1 = a_{-2}\Phi^{-1} + \cdots + a_{-n}\Phi^{-n+1} = 0.a_{-2} \ldots a_{-n} \qquad (7.9)$$

is the *fractional part* of the product (7.8).

Thus, it follows from this consideration that after the first multiplication of the *initial fractional number* (7.7) by base $\Phi$, the integer part of the product (7.8) is the binary numeral $a_{-1}$ of the *"golden"* representation of the *fractional number* (7.7).

By multiplying the fractional number (7.9) by base $\Phi$, we get the following result:

$$A_1 \times \Phi = a_{-2} + a_{-3}\Phi^{-1} + \cdots + a_{-n}\Phi^{-n+2} = a_{-2}.a_{-3} \ldots a_{-n}.$$

$$(7.10)$$

The analysis of (7.10) shows that the second multiplication by base $\Phi$ leads us to the binary numeral $a_{-2}$ of the *"golden"* *representation* of the initial fractional number (7.7).

By continuing the multiplication process $n$ times, we get the "golden" representation of the fractional number $A$.

**Example 7.1.** Convert the decimal fraction $\frac{1}{2}$ into its *"golden"* *representation* in Bergman's system.

**Solution.**

**First multiplication:**

$$\frac{1}{2} \times \Phi = \frac{1}{2} \times \frac{1 + \sqrt{5}}{2} = \frac{1 + \sqrt{5}}{4} \approx 0.809. \qquad (7.11)$$

Since the integer part of the fractional number (7.11) is equal to 0, it follows from this that the first *"golden" binary numeral* of the decimal fraction $\frac{1}{2}$ in *Bergman's system* (8.1) is equal to $a_{-1} = 0$.

**Second multiplication:**

$$\left(\frac{1}{2} \times \Phi\right) \times \Phi = \frac{1}{2} \times \Phi^2 = \frac{1}{2} \times \frac{3 + \sqrt{5}}{2} = \frac{3 + \sqrt{5}}{4} \approx 1.309.$$

$$(7.12)$$

Since the integer part of the obtained product (7.12) is equal to 1, then $a_{-2} = 1$. It follows from this result that before the third multiplication, it is necessary to subtract number 1 from number (7.12):

$$\frac{3 + \sqrt{5}}{4} - 1 = \frac{\sqrt{5} - 1}{4} = \frac{1}{2} \times \Phi^{-1} \approx 0.309.$$

**Third multiplication:**

$$\left(\frac{1}{2} \times \Phi^{-1}\right) \times \Phi^1 = \frac{1}{2} = 0.5.$$

As a result of the third multiplication, we obtained the fractional number 0.5. This means that the *"golden"* bit $a_{-3}$ of the decimal fraction $\frac{1}{2}$ in *Bergman's system* (8.1) is equal to $a_{-3} = 0$.

After the third multiplication, we came to the initial fractional number $\frac{1}{2}$. It follows from this fact that the further multiplication will lead to repetition of the obtained "golden" numerals $a_{-1} = 0$, $a_{-2} = 1$, $a_{-3} = 0$. Hence, the *"golden" representation* of the decimal fraction $\frac{1}{2}$ in *Bergman's system* (8.1) has the form of the following periodic fraction:

$$\frac{1}{2} = 0.010010010\ldots.$$

### 7.2.3. Conversion of integers

The analytic method for the obtaining of the *"golden" representation* of integers reduces to the sequential comparison of the initial number $N$ and the remainders arising here with the powers of the golden ratio.

**Example 7.2.** Convert integer number 4 into the *"golden"* *representation* in *Bergman's system*.

**Solution.** By using the formula $\Phi^n = \frac{L_n + F_n \sqrt{5}}{2}$ [11], we can get the decimal equivalents (D.E.) of the powers of the *golden ratio* (see Table 7.2).

Table 7.2. Decimal equivalents for the powers of the golden ratio.

| $n$ | 3 | 2 | 1 | 0 | $-1$ | $-2$ | $-3$ | $-4$ |
|------|------|------|------|------|------|------|------|------|
| $\Phi^n$ | $\frac{4+2\sqrt{5}}{2}$ | $\frac{3+\sqrt{5}}{2}$ | $\frac{1+\sqrt{5}}{2}$ | 1 | $\frac{-1+\sqrt{5}}{2}$ | $\frac{3-\sqrt{5}}{2}$ | $\frac{-4+2\sqrt{5}}{2}$ | $\frac{7-3\sqrt{5}}{2}$ |
| *D.E.* | 4.236 | 2.618 | 1.618 | 1 | 0.618 | 0.382 | 0.236 | 0.146 |

**The first step of the conversion:** By comparing integer 4 with the D.E. of the golden ratio powers (Table 7.2), we find the pair of the neighboring powers $\Phi^3 = \frac{4+2\sqrt{5}}{2} = 4.236$ and $\Phi^2 = \frac{3+\sqrt{5}}{2} = 2.618$, which are connected with number 4 by the following non-equalities:

$$\frac{3+\sqrt{5}}{2} = 2.618 \leq 4 < \frac{4+2\sqrt{5}}{2} = 4.236. \tag{7.13}$$

It follows from (7.13) that the binary numeral of the second digit of the *"golden"* *representation* of number 4 in *Bergman's system* is equal to $a_2 = 1$.

**The second step of the conversion:** Represent number 4 as follows:

$$4 = \frac{3+\sqrt{5}}{2} + r_1. \tag{7.14}$$

Let's calculate the value of the remainder $r_1$ as follows:

$$r_1 = 4 - \frac{3+\sqrt{5}}{2} = \frac{5-\sqrt{5}}{2} = 1.382. \tag{7.15}$$

By comparing the remainder (7.15) to the golden ratio powers (Table 7.2), we can find the next pair of the golden ratio powers, $\Phi^1 = \frac{1+\sqrt{5}}{2} = 1.618$ and $\Phi^0 = 1$ in *Bergman's system*, which are connected with the reminder $r_1 = 1.382$ by the following non-equality:

$$1 \leq 1.382 < \frac{1+\sqrt{5}}{2} = 1.618. \tag{7.16}$$

It follows from (7.16) that the binary numeral of the zeroth digit of the "golden" representation of number 4 in *Bergman's system* is equal to $a_0 = 1$.

**The third step of the conversion:** Let's represent the first remainder $r_1 = 1.382$ as follows:

$$r_1 = \frac{5 - \sqrt{5}}{2} = 1 + r_2, \qquad (7.17)$$

where the second remainder $r_2$ is equal to:

$$r_2 = r_1 - 1 = \frac{5 - \sqrt{5}}{2} - 1 = \frac{3 - \sqrt{5}}{2} = 0.382. \qquad (7.18)$$

By comparing the second remainder (7.18) to the *golden ratio powers* (Table 7.2), we can find the next pair of the neighboring *golden ratio powers*, $\Phi^{-1} = \frac{-1+\sqrt{5}}{2} = 0.618$ and $\Phi^{-2} = \frac{3-\sqrt{5}}{2} = 0.382$, which are connected with the second remainder $r_2 = 0.382$ by the following non-equality:

$$\frac{3 - \sqrt{5}}{2} = 0.382 \le 0.382 < \frac{-1 + \sqrt{5}}{2} = 0.618. \qquad (7.19)$$

It follows from (7.19) that the *golden numeral $a_{-2}$* of the *"golden"* *representation* of number 4 in *Bergman's system* is equal to $a_{-2} = 1$.

**The fourth step of conversion:** Let's represent the second remainder $r_2 = \frac{3-\sqrt{5}}{2} = 0.382$ as follows:

$$r_2 = \frac{3 - \sqrt{5}}{2} = \frac{3 - \sqrt{5}}{2} + r_3, \qquad (7.20)$$

where the third remainder $r_3$ is equal to

$$r_3 = r_2 - \frac{3 - \sqrt{5}}{2} = 0. \qquad (7.21)$$

Since the third remainder $r_3 = 0$, then the conversion process is over and the *conversion result* in *Bergman's system* (8.1) is as follows:

$$4 = 101.01. \qquad (7.22)$$

Note that Examples 7.1 and 7.2 are the basis for the designing of the computer algorithm of the number conversion to the *"golden"* *representation*.

It is important to emphasize that the algorithm of the number conversion of integer number $N$ to its *"golden"* *representation* in *Bergman's system* (4.1) is similar to the algorithm of the number conversion of $N$ to its binary representation in the *classical binary system*.

## 7.3. The "Golden" Arithmetic

### 7.3.1. The "golden" multiplicative and additive properties

There are the following important properties, which connect the digit weights of the codes (7.6):

$$\Phi_p^i = \Phi_p \times \Phi_p^{i-1} \text{ (the "golden" } \textit{multiplicative} \text{ property)}, \quad (7.23)$$

$$\Phi_p^i = \Phi_p^{i-1} + \Phi_p^{i-p-1} \text{ (the "golden" } \textit{additive} \text{ property)}, \quad (7.24)$$

where $p = 0, 1, 2, 3, \ldots$ and $i = 0, \pm 1, \pm 2, \pm 3, \ldots$.

The properties (7.23) and (7.24) are the basis of the *"golden" arithmetic* in the *codes of the golden p-proportions* (7.6).

From the point of view of the simplicity of the technical implementation, *Bergman's system* is the most interesting; it is a special case ($p = 1$) of the *codes of the golden p-proportions* (7.6). For the case $p = 1$, the *"golden" properties* (7.23) and (7.24) take the following forms:

$$\Phi^i = \Phi \times \Phi^{i-1} \text{ (the "golden" } \textit{multiplicative} \text{ property, } p = 1),$$
$$(7.25)$$

$$\Phi^i = \Phi^{i-1} + \Phi^{i-2} \text{ (the "golden" } \textit{additive} \text{ property, } p = 1),$$
$$(7.26)$$

where $i = 0, \pm 1, \pm 2, \pm 3, \ldots$ and $\Phi = \frac{1+\sqrt{5}}{2}$ (the golden ratio).

The *"golden" identities* (7.25) and (7.26) underlie the *"golden" arithmetic* for *Bergman's system* (8.1) $A = \sum_i a_i \Phi^i$ [19].

### 7.3.2. The "golden" summation and subtraction

Let's compare the *"golden" additive properties* (7.24) and (7.26) to the recurrent relation $F_p(n) = F_p(n-1) + F_p(n-p-1)$ for the *Fibonacci p-numbers* and the recurrent relation $F_n = F_{n-1} + F_{n-2}$ for the *classical Fibonacci numbers*, respectively (see Table 7.3).

Table 7.3. Comparison of the "golden" additive properties to the recurrent relations for the Fibonacci $p$-numbers and the classic Fibonacci numbers ($p = 1$).

| $p$ | "Golden" additive properties | Recurrent relations for the Fibonacci $p$-numbers and the classical Fibonacci numbers |
|---|---|---|
| $p \geq 0$ | $\Phi_p^i = \Phi_p^{i-1} + \Phi_p^{i-p-1}$ | $F_p(i) = F_p(i-1) + F_p(i-p-1)$ |
| $p = 1$ | $\Phi^i = \Phi^{i-1} + \Phi^{i-2}$ | $F_n = F_{n-1} + F_{n-2}$ |

Table 7.3 shows that the *"golden" additive property* (7.24) $\Phi_p^i = \Phi_p^{i-1} + \Phi_p^{i-p-1}$ and the recurrent relation $F_p(i) = F_p(i-1) + F_p(i-p-1)$ for the *Fibonacci p-numbers* as well as the *"golden" additive property* (7.26) $\Phi^i = \Phi^{i-1} + \Phi^{i-2}$ and the recurrent relation $F_n = F_{n-1} + F_{n-2}$ for the classical Fibonacci numbers are similar in their mathematical structures.

As shown above, the recurrent relations for the *Fibonacci p-numbers* $F_p(i) = F_p(i-1) + F_p(i-p-1)$ and the *classical Fibonacci numbers* $F_n = F_{n-1} + F_{n-2}$ are the basis of the original *Fibonacci arithmetical operations*, in particular, the *Fibonacci summation* and *subtraction*, based on the four *basic micro-operations*. It follows from Table 7.3 that the same *basic micro-operations* can be used to fulfill the arithmetic operations of the *"golden" summation and subtraction.*

We had earlier introduced the following codes for the representation of natural numbers: $\Phi$-*code* (4.9) $N = \sum_i a_i \Phi^i$, *F-code* (4.36) $N = \sum_i a_i F_{i+1}$, and *L-code* (4.40) $N = \sum_i a_i L_{i+1}$ (see Chapter 4).

Let's now consider the following examples of the "golden" summation and subtraction of the natural numbers represented in the $\Phi$-*code* (4.9), the *F-code* (4.36) and the *L-code* (4.40).

First of all, let's recall that all the codes (4.9), (4.36), and (4.40) are the distinct *"golden" codes* of one and the same natural number $N$ in *Bergman's system* (4.1), and the bits $a_i \in \{0,1\}$ ($i = 0, \pm 1, \pm 2, \pm 3, \ldots$) for all the codes (4.9), (4.36) and (4.40) are coincident.

Let's note that the codes (4.9), (4.36) and (4.40) are from a very effective method of the error detection, based on the so-called $Z$- and $D$-properties, according to Theorems 4.2 and 4.3.

Let's recall that for the case $i = 0, \pm 1, \pm 2, \pm 3, \ldots$ the $Z$- and $D$-properties are as follows:

**Z-property:**

$$For\ any\ N = \sum_i a_i \Phi^i after\ substitution\ F_i \to \Phi^i,$$
$$we\ have\ \sum_i a_i F_i \equiv 0 \tag{7.27}$$

**D-property:**

$$For\ any\ N = \sum_i a_i \Phi^i\ after\ substitution\ L_i \to \Phi^i,$$
$$we\ have\ \sum_i a_i L_i \equiv 2N \tag{7.28}$$

Let's now consider the following examples of the *"golden"* *summation and subtraction* of natural numbers based on the *basic micro-operations* (6.93).

**Example 7.2.** Let's summarize the natural numbers $5 + 4$ represented in the $\Phi$-*code* (4.9).

**Solution.**

**The first stage:** Let's represent the natural numbers $N_1 = 5$ and $N_2 = 4$ in the $\Phi$- and $F$-codes by using the MINIMAL FORM (see rows 2 and 3 as well as 5 and 6 of Table 7.4).

Table 7.4. "Golden" representations of the numbers $N_1 = 5$ and $N_2 = 4$ in the $\Phi$- and $F$-codes.

| $i$ | $\to$ | 3 | 2 | 1 | 0. | $-1$ | $-2$ | $-3$ | $-4$ |
|---|---|---|---|---|---|---|---|---|---|
| $\Phi^i$ | $\to$ | $\Phi^3$ | $\Phi^2$ | $\Phi^1$ | $\Phi^0$ | $\Phi^{-1}$ | $\Phi^{-2}$ | $\Phi^{-3}$ | $\Phi^{-4}$ |
| $F_i$ | $\to$ | 2 | 1 | 1 | 0. | 1 | $-1$ | 2 | $-3$ |
| $\downarrow N/F_{i+1}$ | $\to$ | 3 | 2 | 1 | 1. | 0 | 1 | $-1$ | 2 |
| $N_1 = 5$ | $=$ | 1 | 0 | 0 | 0. | 1 | 0 | 0 | 1 |
| $N_2 = 4$ | $=$ | 0 | 1 | 0 | 1. | 0 | 1 | 0 | 0 |

250 Mathematics of Harmony as New Interdisciplinary Direction — Vol. II

For the representation of numbers 5 and 4 in the $\Phi$-code, we use rows 2 (digit weights), 5 and 6 (MINIMAL FORMS of numbers 5 and 4).

**The $\Phi$-code of number $N_1 = 5$:** According to the $\Phi$-*code* interpretation (see the rows 3 and 5 of Table 7.4), number $N_1 = 5$ is equal to the following sum:

$$N_1 = (1 \times \Phi^3) + (0 \times \Phi^2) + (0 \times \Phi^1) + (0 \times \Phi^0) + (1 \times \Phi^{-1})$$
$$+ (0 \times \Phi^{-2}) + (0 \times \Phi^{-3}) + (1 \times \Phi^{-4}). \tag{7.29}$$

By using Table 7.2 (the decimal equivalents for the powers of the golden ratio), we can rewrite the sum (7.29) as follows:

$$N_1 = \left(1 \times \frac{4+2\sqrt{5}}{2}\right) + \left(0 \times \frac{3+\sqrt{5}}{2}\right) + \left(0 \times \frac{1+\sqrt{5}}{2}\right)$$

$$+ (0 \times 1) + \left(1 \times \frac{-1+\sqrt{5}}{2}\right)$$

$$+ \left(0 \times \frac{3-\sqrt{5}}{2}\right) + \left(0 \times \frac{4-2\sqrt{5}}{2}\right) + \left(1 \times \frac{7-3\sqrt{5}}{2}\right)$$

$$= \frac{4+2\sqrt{5}}{2} + \frac{-1+\sqrt{5}}{2} + \frac{7-3\sqrt{5}}{2} = 5. \tag{7.30}$$

**The $\Phi$-code interpretation of number $N_2 = 4$:** The similar reasoning and conclusions for the $\Phi$-code interpretation can be made for number 4.

| $n$ | 3 | 2 | 1 | 0 | $-1$ | $-2$ | $-3$ | $-4$ |
|---|---|---|---|---|---|---|---|---|
| $\Phi^n$ | $\frac{4+2\sqrt{5}}{2}$ | $\frac{3+\sqrt{5}}{2}$ | $\frac{1+\sqrt{5}}{2}$ | 1 | $\frac{-1+\sqrt{5}}{2}$ | $\frac{3-\sqrt{5}}{2}$ | $\frac{-4+2\sqrt{5}}{2}$ | $\frac{7-3\sqrt{5}}{2}$ |
| $D.E.$ | 4.236 | 2.618 | 1.618 | 1 | 0.618 | 0.382 | 0.236 | 0.146 |

**The $F$-code interpretation of number $N_1 = 5$:** According to the $F$-*code* interpretation (see the rows 3 and 5 of Table 7.4, number

Table 7.5. The micro-operations of *replacement*.

| $i$ | $\rightarrow$ | 3 | 2 | 1 | 0 | 1 | $-1$ | 2 | $-3$ |
|---|---|---|---|---|---|---|---|---|---|
| $\downarrow N/\Phi^i$ | $\rightarrow$ | $\Phi^3$ | $\Phi^2$ | $\Phi^1$ | $\Phi^0$ | $\Phi^{-1}$ | $\Phi^{-2}$ | $\Phi^{-3}$ | $\Phi^{-4}$ |
| $N_1 = 5$ | $=$ | 1 | 0 | 0 | 0. | 1 | 0 | 0 | 1 |
| $+$ | | $\downarrow$ | | | | $\downarrow$ | | | $\downarrow$ |
| $N_2 = 4$ | $=$ | 0 | 1 | 0 | 1 | 0 | 1 | 0 | 0 |
| $S = 5+4$ | $=$ | 1 | 1 | 0 | 1. | 1 | 1 | 0 | 1 |

$N_1 = 5$ is equal to the following sum:

$$N_1 = (1 \times 3) + (0 \times 2) + (0 \times 1) + (0 \times 1) + (1 \times 0) + (1 \times 1)$$
$$+ 0 \times (-1) + (1 \times 2) = 3 + 2 = 5. \tag{7.31}$$

By checking the *"golden" representation* of number $N_1 = 5$ by the MINIMAL FORM (the two bits of 1 together don't meet), we see that the *"golden" representation* of number $N_1 = 5$ is correct.

By checking the "golden" representation of number $N_1 = 5$ by the *Z-property* (see row 3 of Table 7.4), we come to the following result:

$$1 \times 2 + 0 \times 1 + 0 \times 1 + 0 \times 0 + 1 \times 1 + 0 \times (-1) + 0 \times 2 + 1 \times (-3) = 2 + 1 - 3 = 0.$$

This means that the *"golden" representation* of number $N_1 = 5$ is correct.

**The F-code interpretation of number $N_2=4$:** Similar reasoning and conclusions can be made for the *F-code interpretation* of number 4.

**The second stage:** Let's fulfill the micro-operations of the *replacement* (Table 7.5).

Let's reduce the summation result $S = 5 + 4$ to the MINIMAL FORM:

$$S = 5 + 4 = \boxed{011}\,\boxed{01.1}\,101 = 10010.0101. \tag{7.32}$$

Let's consider the *F-code interpretation* of the MINIMAL FORM of the summation result (7.32) and its checking by the MINIMAL FORM and by the *Z-property* (Table 7.6).

Table 7.6. The $F$-code interpretation and checking of the summation result.

| $i$ | $\rightarrow$ | 4 | 3 | 2 | 1 | 0 | $-1$ | $-2$ | $-3$ | $-4$ |
|---|---|---|---|---|---|---|---|---|---|---|
| $\Phi^i$ | $\rightarrow$ | $\Phi^4$ | $\Phi^3$ | $\Phi^2$ | $\Phi^1$ | $\Phi^0$ | $\Phi^{-1}$ | $\Phi^{-2}$ | $\Phi^{-3}$ | $\Phi^{-4}$ |
| $F_i$ | $\rightarrow$ | 3 | 2 | 1 | 1 | 0 | 1 | $-1$ | 2 | $-3$ |
| $\downarrow N/F_{i+1}$ | $\rightarrow$ | 5 | 3 | 2 | 1 | 1 | 0 | 1 | $-1$ | 2 |
| $S = 5 + 4$ | $=$ | 1 | 0 | 0 | 1 | 0. | 0 | 1 | 0 | 1 |

According to the *F-code interpretation* (see rows 4 and 5 of Table 7.6), the sum $S = 5 + 4$ is as follows:

$$S = (1 \times 5) + (0 \times 3) + (0 \times 2) + (1 \times 1) + (0 \times 1) + (0 \times 0)$$
$$+ (1 \times 1) + [0 \times (-1)] + (1 \times 2) = 5 + 1 + 1 + 2 = 9.$$

By checking the *"golden"* representation of the sum $S = 9$ by the MINIMAL FORM, we conclude that the *"golden"* representation of the number $S = 9$ is correct (the two bits of 1 together don't meet).

By checking the *"golden"* representation of the sum $S = 9$ by the *Z-property* (see rows 3 and 5 of Table 7.6), we get the following result:

$$(1 \times 3) + (0 \times 2) + (0 \times 1) + (1 \times 1) + (0 \times 1)$$
$$+ (0 \times 0) + (0 \times 1) + [1 \times (-1)] + (0 \times 2) + [1 \times (-3)]$$
$$= 3 + 1 - 1 - 3 = 0.$$

This means that the *"golden"* representation of the sum $S = 5 + 4$ is correct.

**Example 7.3.** Subtract the number of 11 from number 3 in *Bergman's system.*

**Solution.** Let's represent numbers 3 and 11 in the MINIMAL FORM of the *F-code*:

$$3 = 00100.0100 \quad \text{and} \quad 11 = 10101.0101 \tag{7.33}$$

and check them by the *Z-property* (Table 7.7).

Table 7.7. Representation of numbers 3 and 11 in the MINIMAL FORMS.

| $i$ | $\rightarrow$ | 4 | 3 | 2 | 1 | 0 | -1 | -2 | -3 | -4 |
|---|---|---|---|---|---|---|---|---|---|---|
| $\Phi^i$ | $\rightarrow$ | $\Phi^4$ | $\Phi^3$ | $\Phi^2$ | $\Phi^1$ | $\Phi^0$ | $\Phi^{-1}$ | $\Phi^{-2}$ | $\Phi^{-3}$ | $\Phi^{-4}$ |
| $F_i$ | $\rightarrow$ | 3 | 2 | 1 | 1 | 0 | 1 | -1 | 2 | -3 |
| $\downarrow N/F_{i+1}$ | $\rightarrow$ | 5 | 3 | 2 | 1 | 1 | 0 | 1 | -1 | 2 |
| $N_1 = 3$ | = | 0 | 0 | 1 | 0 | 0. | 0 | 1 | 0 | 0 |
| $N_1 = 11$ | = | 1 | 0 | 1 | 0 | 1. | 0 | 1 | 0 | 1 |

**Number $N_1=3$:** According to the *F-code interpretation* (see rows 4 and 5), the number $N_1 = 3$ is equal to the following sum:

$$N_1 = 0 \times 5 + 0 \times 3 + 1 \times 2 + 0 \times 1 + 0 \times 1 + 0 \times 0$$

$$+ 1 \times 1 + 0 \times (-1) + 0 \times 2 = 2 + 1 = 3.$$

By checking the *"golden" representation* of number $N_1 = 3$ by the *Z-property*, we get the following result:

$$N_2 = 0 \times 2 + 1 \times 1 + 0 \times 1 + 1 \times 0 + 0 \times 0 + 0 \times 1 + 1 \times (-1)$$

$$+ 0 \times 2 + 0 \times (-3) = 1 - 1 = 0.$$

This means that the "golden" representation of number $N_1 = 3$ is correct.

**Number $N_2=11$:** According to the *F-code interpretation*, the number $N_2 = 11$ is equal to the following sum:

$$N_2 = 1 \times 5 + 0 \times 3 + 1 \times 2 + 0 \times 1 + 1 \times 1$$

$$+ 0 \times 0 + 1 \times 1 + 0 \times (-1) + 1 \times 2$$

$$= 5 + 2 + 1 + 1 + 2 = 11.$$

By checking the "golden" representation of number $N_2 = 11$ by the *Z-property*, we get the following result:

$$N_2 = 1 \times 3 + 0 \times 2 + 1 \times 1 + 1 \times 0 + 0 \times 1 + 1 \times 0$$

$$+ 0 \times 1 + 1 \times (-1) + 0 \times 2 + 1 \times (-3) = 3 + 1 - 1 - 3 = 0.$$

This means that the *"golden" representation* of number $N_2 = 11$ is correct.

Table 7.8. The micro-operations of *absorption*.

| $i$ | $\rightarrow$ | 4 | 3 | 2 | 1 | 0 | -1 | -2 | -3 | -4 |
|---|---|---|---|---|---|---|---|---|---|---|
| $\downarrow N/\Phi^i$ | $\rightarrow$ | $\Phi^4$ | $\Phi^3$ | $\Phi^2$ | $\Phi^1$ | $\Phi^0$ | $\Phi^{-1}$ | $\Phi^{-2}$ | $\Phi^{-3}$ | $\Phi^{-4}$ |
| $N_1 = 3$ | $=$ | 0 | 0 | 1 | 0 | 0. | 0 | 1 | 0 | 0 |
| — | $\rightarrow$ | | | $\updownarrow$ | | | | $\updownarrow$ | | |
| $N_2 = 11$ | $=$ | 1 | 0 | 1 | 0 | 1. | 0 | 1 | 0 | 1 |
| $D = 3 - 11$ | $=$ | 1 | 0 | 0 | 0 | 1. | 0 | 0 | 0 | 1 |

Let's now fulfill the micro-operations of *absorption* (Table 7.8).

Since the subtraction result $D = 3 - 11$ is in the bottom register, then the result $D$ has the sign "–".

## 7.4. Representation of Numbers with the Floating Point

Let's now compare the *codes of the golden p-proportions* (7.6) to the *Fibonacci p-codes* (6.1). These codes are similar in their mathematical structures. However, the following distinctions exist between them:

(1) First of all, it is necessary to note that the *Fibonacci p-codes* (6.1) are intended for the representation of integers and demand less digits for the representation of one and the same range of the number representation in comparison to the *codes of the golden p-proportions* (7.6). For example, the decimal number 10 needs for its representation in *Bergman's system* the nine-digit "*golden*" *representation*:

$$10_{10} = 10100.0101$$

$$= 1 \times \Phi^4 + 0 \times \Phi^3 + 1 \times \Phi^2 + 0 \times \Phi^1 + 0 \times \Phi^0$$

$$+ 0 \times \Phi^{-1} + 1 \times \Phi^{-2} + 0 \times \Phi^{-3} + 1 \times \Phi^{-4}$$

$$= \Phi^4 + \Phi^2 + \Phi^{-2} + \Phi^{-4}$$

$$= \frac{7 + 3\sqrt{5}}{2} + \frac{3 + \sqrt{5}}{2} + \frac{3 - \sqrt{5}}{2} + \frac{7 - 3\sqrt{5}}{2} = 7 + 3.$$

$$(7.34)$$

However, for the *Fibonacci 1-code*, we need only six binary digits for the representation of the same number 10:

$$10_{10} = 100\,100_F = 1 \times 8 + 1 \times 2 = 8 + 2.$$

Thus, we can give the preference to the *Fibonacci p-codes* for the *Fibonacci representation* of integers.

(2) The next distinction between the *Fibonacci p-codes* (10.1)

$$N = a_n F_p(n) + a_{n-1} F_p(n-1) + \cdots + a_i F_p(i) + \cdots + a_n F_p(1)$$

and the *codes of the golden p-proportions* (7.6) $A = \sum_i a_i \Phi_p^i$ is connected with digit weights. The digit weights of the *codes of the golden p-proportions* (7.6) form the geometric progression (7.5). This fact gives a possibility to fulfill the left and right shifts of the *"golden"* representations. This corresponds to the *multiplication* and *division* of the initial number by base $\Phi_p$ (the *golden p-ratio*). The *"shift-property"* of the *golden p-ratio codes* gives a possibility to represent numbers with the *floating point*. In fact, the decimal number 10 can be represented in *Bergman's system* with floating point as follows:

$$10_{10} = 10100.0101_\Phi = 0.101000101 \times \Phi^5. \qquad (7.35)$$

The *"golden"* representation with floating point (7.35) consists of two parts. The first part is the *"golden"* mantissa of the decimal number 10,

$$m(10) = 0.101000101, \qquad (7.36)$$

and the second part is the *golden ratio power* $\Phi^5$. Number 5 is called the *"golden"* exponent of number 10 in *Bergman's system*.

## 7.5. The "Golden" Multiplication

In general, the *"golden"* multiplication is based on the following trivial identity for the *golden p-ratio powers*:

$$\Phi_p^n \times \Phi_p^m = \Phi_p^{n+m}. \qquad (7.37)$$

Table 7.9 follows from (7.37) for the *"golden"* multiplication, which is true for the all *codes of the golden p-proportions* (7.6).

Table 7.9. Table of the "golden" multiplication.

| $0 \times 0$ | $=$ | 0 |
|---|---|---|
| $0 \times 1$ | $=$ | 0 |
| $1 \times 0$ | $=$ | 0 |
| $1 \times 1$ | $=$ | 1 |

We can see that Table 7.9 for the *"golden" multiplication* coincides with the multiplication table used in the classical *binary arithmetic*. This means that the *"golden" multiplication* reduces to the *classical binary multiplication*, i.e., to the following rules:

**Rule 1.** To form the partial products in accordance with Table 7.9.

**Rule 2.** To summarize the partial products in accordance with the rule of the *"golden" summation*.

**Example 7.4.** Multiply the *"golden" fractions* $A = 0.010010$ and $B = 0.001010$ in *Bergman's system*.

**Solution 1.** Let's represent the *"golden" fractional numbers* $A = 0.010010$ and $B = 0.001010$ in the form with *floating point*:

$$A = 010010 \times \Phi^{-6},$$
$$B = 001010 \times \Phi^{-6}.$$

This means that the *"golden" mantissas* and the *"golden" exponents* of numbers $A$ and $B$ are equal to the following, respectively:

$$m(A) = 010010; \ e(A) = -6 \text{ and } m(B) = 001010; \ e(B) = -6.$$

**2.** Let's multiply the *"golden" mantissas* $m(A)$ and $m(B)$:

$$
\begin{array}{r}
m(A) = 0\,1\,0\,0\,1\,0 \\
m(B) = 0\,0\,1\,0\,1\,0 \\
\hline
0\,0\,0\,0\,0\,0 \\
0 \quad 1\,0\,0\,1\,0 \\
0 \quad 0 \quad 0\,0\,0\,0 \\
0\,1 \quad 0 \quad 0\,1\,0 \\
0\,0\,0 \quad 0 \quad 0\,0 \\
0\,0\,0\,0 \quad 0 \quad 0 \\
\hline
0\,0\,0\,1 \quad 0 \quad 1\,1\,0\,1\,0\,0 \\
\end{array}
$$

$$\underbrace{\qquad\qquad\qquad\qquad}_{m(A) \times m(B)}$$

**3.** Let's reduce the product $m(A) \times m(B)$ to the MINIMAL FORM:

$$m(A) \times m(B) = 0001\boxed{011}0100 = 00\boxed{011}000100 = 00100000100.$$

**4.** Let's summarize the *"golden"* *exponents*:

$$e(A) + e(B) = (-6) + (-6) = -12.$$

**5.** Let's represent the result of the *"golden"* *multiplication* $A \times B$ in the form with the *floating point*:

$$A \times B = 00100000100 \times \Phi^{-12}.$$

## 7.6. The "Golden" Division

For the creation of the rules of the *"golden"* *division,* we can use an analogy between the *classical binary division* and the *"golden"* *division* for the case of *Bergman's system.* As well known, the *classical binary division* reduces to the *shift* of the *divisor* and its *comparison* with the *divisible* or with the *intermediate divisible.* If the *divisible* or the *intermediate divisible* is more than the *shifted divisor,* then we write the *binary numeral* of 1 to the corresponding digit of the *quotient,* in the opposite case, we write the *binary numeral* of 0. Then, the *shifted divisor* is subtracted from the *divisible* or the *intermediate divisible.* One may see that a similar procedure also underlies the *"golden"* *division.* Since the comparison of numbers is fulfilled over the numbers, represented in the MINIMAL FORM, it follows from this consideration that the peculiarity of the *"golden"* *division* consists in the fact that the *"golden"* *divisible* and all the *intermediate* *"golden"* *divisible* reduce to the MINIMAL FORM at every step of the *"golden"* *division.*

Let's now demonstrate the *"golden"* *division* on the example of *Bergman's system.*

**Example 7.5.** Divide the *"golden"* *representation* of number $5 = 1000.1001$ (the *"golden"* *divisible*) by the *"golden"* *representation* of number $10 = 10100.0100$ (the *"golden"* *divisor*) in *Bergman's system.*

**Solution.** Let's represent the given numbers 5 and 10 in the form with the *floating point*:

$$m(A) = 5 \times \Phi^4 = 10001001; \ e(A) = 4, \tag{7.38}$$

$$m(B) = 10 \times \Phi^4 = 101000101; \ e(B) = 4. \tag{7.39}$$

Because their *"golden"* *exponents* are equal, that is, $e(A) = e(B) = 4$, we can write

$$5 : 10 = m(A) : m(B),$$

that is, the *"golden"* *division* of the initial numbers 5:10 reduces to the *"golden"* *division* of their *"golden"* *mantissas* $m(A) : m(B)$.

Let's now divide the *"golden"* *mantissa* $m(A)$ by the *"golden"* *mantissa* $m(B)$:

1. If we compare the *"golden"* *mantissa* $m(A) = 10001001$ on the *"golden"* *mantissa* $m(B) = 101000101$, we find that $m(A) < m(B)$. This means that the result of the *"golden"* *division* is a proper fraction, that is, the binary numeral of the zeroth digit of the *"golden"* *quotient* $Q$ is equal to $a_0 = 0$.

2. If we shift the *"golden"* *mantissa* $m(A) = 10001001$ by one digit to the left, we get the following *"golden"* *representation*:

$$A_1 = 100010010. \tag{7.40}$$

3. If we compare the *"golden"* *representation* (7.40) to the *"golden"* *mantissa* $m(B) = 101000101$, we find that $A_1 < m(B)$. This means that the binary numeral of the next digit of the *"golden"* *quotient* $Q$ is equal to $a_{-1} = 0$.

4. By shifting the *"golden"* *representation* (7.40) by one digit to the left, we get the following *"golden"* *representation*:

$$A_2 = 1000100100. \tag{7.41}$$

5. If we compare the *"golden"* *representation* (7.40) to the *"golden"* *mantissa* $m(B) = 101000101$, we find that $A_2 \geq m(B)$. This means that the next binary numeral of the *"golden"* *quotient* $Q$ is equal to $a_{-2} = 1$.

6. As we have obtained the bit $a_{-2} = 1$, we need to subtract the *"golden"* mantissa $m(B) = 101000101$ from the *"golden"* representation $A_2 = 1000100100$. As the outcome, we find the next intermediate *"golden"* representation:

$$A_3 = A_2 - m(B) = 1000100.1. \tag{7.42}$$

A peculiarity of the *"golden"* representation (7.42) consists in the fact that it has the bit of 1 in its *fractional part*.

7. By shifting the *"golden"* representation (7.42) by one digit to the left, we get the new intermediate *"golden"* representation:

$$A_4 = 10001001. \tag{7.43}$$

Note that the *"golden"* representation (7.43) is equal to the *"golden"* mantissa $m(A) = 10001001$. This means that starting from the *"golden"* representation (7.43), the process of the *"golden"* division begins to repeat. Hence, the next binary numerals of the *"golden"* quotient will be equal:

$$a_{-3} = 0, a_{-4} = 0, a_{-5} = 1 \dots .$$

It follows from this consideration that the *"golden"* quotient has the following *"golden"* representation:

$$Q = 0.010010010 \dots , \tag{7.44}$$

that is, for this case of the *"golden"* quotient, $Q = (5:10) = 0.5$ is represented in *Bergman's system* as a *periodic fraction* (7.44).

If we compare the *"golden"* arithmetic with the *Fibonacci arithmetic* and the *classical binary arithmetic*, we can find two peculiar properties of the *"golden"* arithmetic:

1. The rules of the *"golden"* summation and subtraction coincide with the corresponding rules for the *Fibonacci arithmetic*.
2. Similar to the classical binary system, the *codes of the golden p-proportions* (7.6) have the important arithmetical property to represent the *"golden"* representations in the form with the *floating point*.

3. The rules of the *"golden" multiplication and division* coincide with similar rules for the *classical binary arithmetic*.

Thus, the *"golden"* arithmetic is a peculiar synthesis of the *Fibonacci* and *classic binary arithmetic*.

## 7.7. Digital Metrology and a New Theory of Resistive Dividers Based on Golden *p*-Proportion Codes

### 7.7.1. What is metrology?

We can find the answer to this question in the Russian and English Wikipedia. The Russian Wikipedia provides the following:

> *"Metrology is the science about measurement, methods and tools to ensure their unity and ways to achieve the required accuracy. The subject of the metrology is to extract quantitative information about the properties of the metrological objects with the specified accuracy and reliability; the metrological standards are the regulatory framework for this."*

The English Wikipedia gives the following answer to this question:

> *"Metrology is defined by the International Bureau of Weights and Measures as the science of measurement, embracing both experimental and theoretical determinations at any level of the uncertainty in any field of science and technology."*

In this section, we describe the *golden resistive divisors* as the basis of self-correcting analog-to-digital and digital-to-analog converters.

For the first time, this original theory was described in Stakhov's 1978 article "Digital Metrology on the base of the Fibonacci codes and Golden Proportion Codes" [58].

### 7.7.2. A structure of the resistive dividers

In the measurement practice, the so-called *resistive dividers*, intended for the division of electric currents and voltages in the given ratio, are widely used. A structure of the resistive divider is shown in Fig. 7.1.

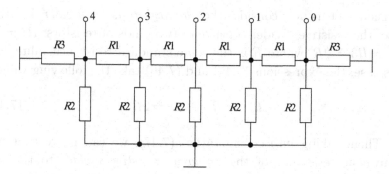

Fig. 7.1. A structure of the resistive divider.

The resistive divider in Fig. 7.1 consists of the "horizontal" resistors of the kind $R1$ and $R3$ and the "vertical" resistors of the kind $R2$. The resistors of the divider are connected between themselves by the "connecting points" 0, 1, 2, 3, 4. Each point connects three resistors, which together form the resistor section. Note that Fig. 7.1 shows the resistive divider, which consists of the five resistor sections. In general, a number of resistor sections can be extended *ad infinitum*.

First of all, we note that the parallel connection of the resistors $R2$ and $R3$ to the right of the "connecting point" 0 and to the left of the "connecting point" 4 can be replaced by the equivalent resistor $R_{e1}$ with the resistance, which can be calculated according to the well-known electric law on the parallel connection of the two resistors $R2$ and $R3$:

$$R_{e1} = \frac{R2 \times R3}{R2 + R3}. \tag{7.45}$$

Then, it is easy to calculate the equivalent resistance of the resistor section to the right of the "connecting point" 1 and to the left of the "connecting point" 3:

$$R_{e2} = R1 + R_{e1}. \tag{7.46}$$

### 7.7.3. The "binary" resistive divider

Depending on the choice of the resistance values of the resistors $R1$, $R2$, $R3$, we can get different coefficients for the current or voltage division. Let us now consider the *"binary" resistive divider.* For this case, the resistive divider consists of two kinds of resistors: $R1 = R$, $R2 = R3 = 2R$, where $R$ is the some standard resistance value. For this case, the expressions (7.45) and (7.46) take the following values:

$$R_{e1} = R; \quad R_{e2} = 2R. \tag{7.47}$$

Then, taking into consideration (7.47), we can prove that the equivalent resistance of the *"binary" resistive divider* to the left or to the right of some "connecting points" 0, 1, 2, 3, 4 is equal to $2R$. This means that the equivalent resistance of the resistive *divider* in the "connecting points" 0, 1, 2, 3, 4 can be calculated as the resistance of the parallel connection of the three resistors of the value $2R$. By using the electric circuit laws, we can calculate the equivalent resistance of the *"binary" resistive divider* in the "connecting points" 0, 1, 2, 3, 4 as follows:

$$R_{e3} = \frac{2}{3}R. \tag{7.48}$$

Let's now connect the generator of the standard electric current $I$ to one of the "connecting points", for example, to the point 2. Then according to Ohm's law, the following electric voltage appears at this point:

$$U = \frac{2}{3}RI. \tag{7.49}$$

Let's now calculate the electrical voltages in the "connecting points" 3 and 1, which are adjacent to point 2. It is easy to show that the voltage transmission coefficient between the adjacent "connecting points" is equal to $\frac{1}{2}$. This means that the *"binary" resistive divider* fits very well with the *binary system* and this fact is the reason for the wide use of the *"binary" resistive divider* (Fig. 7.1) in the modern "binary" digital-to-analog and analog-to-digital converters.

## 7.8. The "Golden" Self-Correcting Digital-to-Analog and Analog-to-Digital Converters

### 7.8.1. The electrical properties of the "golden" resistive dividers

Let's choose the values of the resistors of the *"golden" resistive divider* in Fig. 7.1 as follows:

$$R1 = \Phi_p^{-p}R; \quad R2 = \Phi_p^{p+1}R; \quad R3 = \Phi_p R, \qquad (7.50)$$

where $\Phi_p$ is the *golden p-ratio*, $p \in \{0, 1, 2, 3, \dots\}$.

It is clear that the "golden" resistive divider in Fig. 7.1 sets the infinite number of the different *"golden" resistive dividers* because every $p$ "generates" the new *"golden" resistive divider*. In particular, for the case $p = 0$, the value of the *golden 0-ratio* $\Phi_0 = 2$ and the *"golden" resistive divider* reduces to the classical *"binary" resistive divider* based on the resistors $R \div 2R$.

For the case $p = 1$, the resistors $R1$, $R2$, $R3$ in Fig. 7.1 take the following values:

$$R1 = \Phi^{-1}R = 0.618R; \quad R2 = \Phi^2 R = 2.618R; \quad R3 = \Phi R = 1.618R.$$
$$(7.51)$$

For this choice of the values of the resistors $R1$, $R2$, $R3$, given by (7.50), (7.51), the *"golden" resistive dividers* in Fig. 7.1 have the following unique electrical properties from the mathematical relations of the *golden p-proportions* $\Phi_p$:

$$\Phi_p = 1 + \Phi_p^{-p}, \qquad (7.52)$$
$$\Phi_p^{p+2} = \Phi_p^{p+1} + \Phi_p, \qquad (7.53)$$

which takes the following forms for the cases $p = 0$ ($\Phi_{p=0} = 2$) and $p = 1$ ($\Phi_{p=1} = \Phi = \frac{1+\sqrt{5}}{2} = 1.618$), respectively:

$$p = 0 : 2 = 1 + 1; \quad 2^2 = 2 + 2, \qquad (7.54)$$
$$p = 1 : \Phi = 1 + \Phi^{-1}; \quad \Phi^3 = \Phi^2 + \Phi. \qquad (7.55)$$

By using the identity (7.53), we can deduce the value of the equivalent resistance of the resistor circuit of the *"golden" resistive divider* in Fig. 1.11 in Volume I to the left and to the right from the "connecting points" 0 and 4. In general case of $p$ ($p \geq 1$), the formula (7.45) is written as follows:

$$R_{e1} = \frac{R2 \times R3}{R2 + R3} = \frac{\Phi_p^{p+1} R \times \Phi_p R}{\Phi_p^{p+1} R + \Phi_p R} = \frac{\Phi_p^{p+2} R^2}{(\Phi_p^{p+1} + \Phi_p)R} = R. \quad (7.56)$$

Note that we have simplified the formula (7.56) by using the mathematical identity (7.53).

By using (7.46) and (7.52), we can calculate the value of the equivalent resistance $R_{e2}$ as follows:

$$R_{e2} = \Phi_p^{-p} R + R = (\Phi_p^{-p} + 1)R = \Phi_p R. \quad (7.57)$$

Thus, according to (7.57), the equivalent resistance of the resistive circuit of the *"golden" resistive divider* to the left or to the right of the "connecting points" 0, 1, 2, 3, 4 is equal to $\Phi_p R$, where $\Phi_p$ is the *golden p-proportion*. This fact can be used for the calculation of the equivalent resistance $R_{e3}$ of the *"golden" resistive divider* in the "connecting points" 0, 1, 2, 3, 4. In fact, the equivalent resistance $R_{e3}$ can be calculated as the resistance of the electrical circuit, which consists of the parallel connection of the "vertical" resistor $R2 = \Phi_p^{p+1} R$ and the two "lateral" resistors with the resistance $\Phi_p R$. Since, according to (7.56), the equivalent resistance of the parallel connection of the resistors $R2 = \Phi_p^{p+1} R$ and $R3 = \Phi_p R$ is equal to $R$, then the equivalent resistance $R_{e3}$ of the "golden" resistive divider in each "connecting point" can be calculated by the following formula:

$$R_{e3} = \frac{\Phi_p R \times R}{\Phi_p R + R} = \frac{\Phi_p R^2}{(\Phi_p + 1)R} = \frac{1}{1 + \Phi_p^{-1}} R. \quad (7.58)$$

Note that for the case $p = 0$ (the "binary" resistive divider), $\Phi_{p=0} = 2$ and then the expression (7.58) reduces to (7.48). For the case $p = 1$, the formula (7.58) reduces to the following formula:

$$R_{e3} = \frac{1}{1 + \Phi^{-1}} R = \frac{1}{\Phi} R = \Phi^{-1} R. \tag{7.59}$$

Let's now calculate the voltage transmission coefficient between the adjacent "connecting points" of the "*golden*" *resistive divider*. For this purpose, we connect the generator of the standard electric current $I$ to one of the "connecting points", for example, to point 2. Then, according to Ohm's law, the following electrical voltage appears at this point:

$$U = \frac{1}{1 + \Phi_p^{-1}} RI. \tag{7.60}$$

Note that for the case $p = 0$, we have $\Phi_{p=0} = 2$ and the formula (7.60) reduces to the following formula:

$$U = \frac{1}{1 + 2^{-1}} RI = \frac{1}{1 + \frac{1}{2}} RI = \frac{2}{3} RI, \tag{7.61}$$

which coincides with the formula (7.47) for the "binary" resistive divider.

Let's now calculate the electrical voltage in the adjacent "connecting points" 3 and 1:

1. The voltages in points 3 and 1 can be calculated as a result of linking the voltage $U$, given by (7.60), to the resistive circuit, which consists of the sequential connection of the "*horizontal*" *resistor* $R1 = \Phi_p^{-p} R$ and the resistive circuit with the equivalent resistance $R$. Then, for this case, the electrical current $I$, which appears in the resistive circuit to the left and to the right of the "connecting point" 2, will be equal to

$$I = \frac{U}{R1 + R} = \frac{U}{(\Phi_p^{-p} + 1)R} = \frac{U}{\Phi_p R}. \tag{7.62}$$

If we multiply the electrical current (7.62) by the equivalent resistance $R$, we get the following value of the electrical voltage in the adjacent "connecting points" 3 and 1:

$$\frac{U}{\Phi_p}. \tag{7.63}$$

This means that the voltage transmission coefficient between the adjacent "connecting points" of the "golden" resistive divider in Fig. 7.1 is equal to the reciprocal to the *golden p-proportion* $\Phi_p$!

Thus, the *"golden" resistive divider* in Fig. 7.1, based on the *golden p-proportions*, is quite real electrical circuits. It is clear that the above theory of the *"golden" resistive dividers* [16] can become a new source for the development of the "digital metrology" and self-correcting analog-to-digital and digital-to-analog converters.

### 7.8.2. The "golden" digital-to-analog converters

The electrical circuit of the *"golden" digital-to-analog converters* (DAC), based on the *"golden"* resistive divider in Fig. 7.1, is shown in Fig. 7.2. Note that the *"golden"* DAC in Fig. 7.2 consists of five digits. However, the number of the DAC digits may be increased to some arbitrary $n$ by extending the *"golden" resistive divider* to the left and to the right.

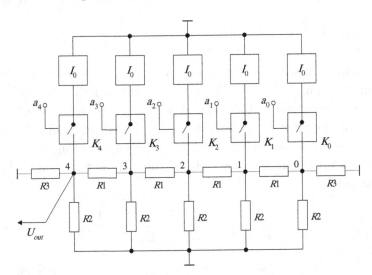

Fig. 7.2. The "golden" DAC.

The "golden" DAC contains five ($n$ in the general case) generators of the standard electrical current $I_0$ and five ($n$ in the general case) electrical current keys $K_0 - K_4$. The key states are controlled by the binary digits of the *golden p-proportion code* $a_4a_3a_2a_1a_0$. For the case $a_i = 1$, the key $K_i$ is closed, for the case $a_i = 0$, the key is open ($i = 0, 1, 2, \ldots, n$).

One can show that the *closed* key $K_i$ results in the following voltage in the $i$th "connecting point" of the resistive divider: $U_i = \beta_p I_0 R$, where $\beta_p = \frac{1}{1+\Phi_p^{-1}}$.

As the potential $U_i$ is passed from the $i$th point to the $(i+1)$th point with the transmission coefficient $\frac{1}{\Phi_p}$, the following voltage appears at the DAC output:

$$U_{out} = \frac{\beta_p I_0 R}{\Phi_p^{n-l-1}} = \frac{\beta_p I_0 R}{\Phi_p^{n-1}} \times \Phi_p^i.$$

By using the superposition principle, it is easy to show that the *golden p-proportion code* $a_{n-1}a_{n-2}\ldots a_0$ results in the following voltage $U_{out}$:

$$U_{out} = B_p \sum_{i=0}^{n-1} a_l \Phi_p^i, \tag{7.64}$$

where $B_p = \frac{\beta_p I_0 R}{\Phi_p^{n-1}}$.

It follows from (7.64) that the electrical circuit in Fig. 7.2 converts the *golden p-proportion code* $a_4a_3a_2a_1a_0$ into the electrical voltage $U_{out}$ with regard to the constant coefficient $B_p$.

### 7.8.3. Checking the "golden" DAC

In the measurement practice, there is a necessity to check up the DAC linearity at the production and operation processes. In the classical binary DAC, the following correlation is used for checking up the DAC linearity:

$$2^n = \sum_{i=0}^{n-1} 2^i + 1.$$

The mathematical properties of the *golden p-proportion codes* (7.6) are providing very wide possibilities for checking up the "golden" DAC linearity. In particular, the linearity checking up of the "golden" DAC, based on the *classical golden ratio* ($p = 1$), reduces to the checking up of the following relations:

$$\Phi^n = \Phi^{n-1} + \Phi^{n-2} = \Phi^{n-1} + \Phi^{n-3} + \Phi^{n-4}$$

$$= \Phi^{n-1} + \Phi^{n-3} + \Phi^{n-5} + \Phi^{n-6}. \tag{7.65}$$

The checking up is fulfilled in the following manner. We have to check that the output voltage of the "golden" DAC in Fig. 7.2 will be constant for the following input *"golden" representations*, which corresponds to (7.65):

$$\boxed{100}\,0000 = 01\boxed{100}\,00 = 0101\boxed{100} = 0101011.$$

Note that the different input *"golden" representations* are formed from the *top "golden" representation* 1000000 by means of the *devolutions* $\boxed{100 \rightarrow 011}$.

### 7.8.4. The "golden" self-correcting ADC

The *Fibonacci code* and the *golden proportion codes* allow the application of the *principle of self-correction* to improve the exactness and metrological stability of ADC. The correction of the *nonlinearity* of transfer function of the *"golden" resistive divider* at the production and operation of ADC was the most important advantage of ADC.

In the Special Design Bureau "Module" of the Vinnitsa Technical University (Ukraine), under Prof. Stakhov's scientific leadership, several modifications of the self-correcting ADC were developed, and a special procedure for the correction of deviations of the digit weights from their standard (ideal) values (the Fibonacci numbers or the golden ratio powers) was realized.

The self-correcting 17-digit ADC, based on the Fibonacci numbers, was one of the best Soviet ADCs, which was designed and produced in the Special Design Bureau "Module" in the 1980s [23] (Fig. 7.3). This ADC was widely used in various metrology

Fig. 7.3. The 17-digit self-correcting ADC.

firms, located in Moscow, Leningrad, Riga, Kiev and other cities of the USSR. For the development of the self-correcting ADC, Alexey Stakhov was awarded with the "Gold Medal" of the Exhibition of Achievements of the USSR Economy (Moscow).

The ADC in Fig. 7.3 had the following technical parameters:

1. The number of digits — 17 (16 digits and one sign digit);
2. Conversion time — 15 $\mu$s;
3. Total error — 0.006%;
4. Linearity error — 0.003%;
5. Frequency range — 25 kHz;
6. Operating temperature range — $20 \pm 30°$C.

The correction system, built in the ADC, allows the correction of the *zero drift*, the *linearity* of the analog-to-digital conversion, which is fulfilled by the traditional methods, and most importantly, the correction of the *deviations* of the digit weights from their standard values (the *Fibonacci numbers* or the *powers of the golden proportion*).

According to the well-known Soviet metrological firms, in that period, the Soviet electronic industry had not produced ADC with such high technical parameters.

The correction procedure was used in the manufacturing process for the automatic adjustment of ADC to the given exactness and then it was periodically repeated depending on the aging of the elements and external factors, for example, temperature changes which lead to the deviation of the digit weights from their standard (ideal) values (the *Fibonacci numbers* or the *powers of the golden proportion*).

# Conclusions

## General Conclusions

(1) The discovery by the *Babylonians of the positional principle of number representation* and then by the *Indians of the decimal numeral system*, based on the Babylonian positional principle, as well as the creation of the *binary, ternary* and other positional numeral systems can rightly be classified as *truly epochal achievements* in the entire history of mathematics that significantly influenced the development of material culture and science, in particular, of *mathematics* and *computer science.*

(2) The scornful attitude towards the *"school"* and *applied* arithmetic and its problems, as well as the absence of a sufficiently serious need for the creation of new numeral systems in computational practice, which during the past centuries was satisfied completely by the *decimal system* (in the educational practice) and in the recent decades by the *binary system* (in computer science), can serve as an explanation to the fact as to why in the number theory due attention wasn't given to numeral systems and why during this time, mathematics did not move far ahead in comparison to the initial period of its development.

The interest for the numeral systems was preserved only in the history of mathematics. Although historians of mathematics claim that it was precisely the count problem that was one of the main stimulating factors for the development of mathematics at the stage of its origin, after the invention of the decimal, binary, and ternary numeral systems, this problem was considered in mathematics as completely solved and the serious mathematical studies of this direction (except historical research about the origin of numeral systems and discussion of the methods of teaching of the decimal system in secondary schools) in mathematics was not conducted.

(3) The situation changed dramatically in the second half of the 20th century after the appearance of modern computers. It was in this area that a great interest was again shown to the *methods of the representation of numbers and new computer arithmetics* [133–136]. Apart from the usage of the *binary system* (von Neumann's principles), already at the initial stage of the "computer era", attempts were made to use the other non-binary numeral systems in computer science. In this respect, the ternary computer "Setun", designed by the Russian engineer Nikolay Brusentsov on the basis of the ternary system, was an excellent example. During this period, the numeral systems with the "exotic" names and properties appeared: the *system of residual classes*, a system with the *complex-valued base, nega-positional, factorial, binomial* numeral systems, etc. [135, 136]. All of them had certain advantages when compared with the binary system and were aimed at improving certain technical characteristics of computers. Some of them became the basis for the design of the new computer projects (the ternary computer "Setun", the computer, based on the *residual classes system*, etc.).

(4) But there is the following interesting historical aspect of this problem. Around four millennia after the Babylonians invented the *positional principle*, a peculiar period of the *"Renaissance"* arose in the field of numeral systems. Thanks to the efforts, first of all, of specialists in the field of computer arithmetic [133–136], mathematics again returned to the period of its origin, when the numeral systems were regarded as one of the main subjects of ancient mathematics (*Babylonian positional principle, Egyptian principle of doubling* as the main principle of the *Egyptian arithmetic*, and so on). But this initial period of mathematics origin, according to many historians of mathematics [50, 133],

is considered as extremely important for the development of numeral systems and mathematics on the whole. It was during this period that the foundations of such important mathematical discipline as the *elementary theory of numbers*, which is rightly called the "queen of mathematics", were laid. But then it's quite reasonable to ask the following question: *could the newest numeral systems (created to meet the utilitarian needs of computer technology) influence the development of the fundamental concepts of the elementary number theory and thus influence not only the development of computer science but also the development of all mathematics in general?* Searching the answer to this question was one of the purposes of Volume II of this three-volume book.

(5) The detailed historical studies on the issue of the origin of the *numeral systems*, presented in Neugebauer's book *Lectures on the History of Ancient Mathematical Sciences* (1934) [134], in the article by Bashmakova and Yushkevich, "the origin of the numeral systems" (1951) [133], in the book by the prominent Russian mathematician academician Kolmogorov *Mathematics in its Historical Development* (1991) [102], led us to a very important conclusion, most clearly expressed in the following citation, taken from the article "Mathematics as a Science about Models" written by the famous Bulgarian mathematician academician Iliev (1972) [137]:

> *"During the first epoch of its development — from antiquity and right up to the discovery of differential and integral calculus — mathematics, by exploring primarily the problem of magnitudes measuring, created Euclidean geometry and the theory of numbers."*

This quotation became the basis for uniting in Volume II the two newest applied mathematical theories that are important for further development of mathematics and computer science: the *algorithmic measurement theory* (Chapters 1 and 2) [16] and the *theory of numeral systems with irrational bases* (Chapters 3–7) [54–62, 65, 70–72, 99, 100].

## List of the Most Important Modern Scientific Results

## Chapters 1 and 2

(6) The main scientific result, from Chapter 1 and Chapter 2 reduces to the creation of the *constructive (algorithmic) measurement theory* [16] based on the *constructive approach* to the conception of *mathematical infinity* (the *potential infinity*). A distinctive feature of the *algorithmic measurement theory* in comparison with the classical mathematical measurement theory, based on the *continuity axioms* (*Eudoxus–Archimedes and Cantor axioms*), is the fact that the *algorithmic measurement theory excludes from the consideration* the *abstraction of the actual infinity*, which is the *cornerstone* of Cantor's theory of infinite sets.

Thanks to the constructive approach to the concept of *mathematical infinity* (the *potential infinity*), in the *algorithmic measurement theory*, it was possible to get the following new scientific results:

(i) Formulation of the concept of the *optimal measurement algorithms* [16];

(ii) Development of the new look at the oldest *Fibonacci weighing problem* (Fibonacci, 13th century) called the *Bachet–Mendeleev problem* in the Russian historical and scientific literature [16];

(iii) Formulation of the *principle of measurement asymmetry* [57];

(iv) Synthesis of the new classes of *optimal measurement algorithms*, in particular, the *binomial* and *Fibonacci* measurement algorithms [6, 16];

(v) Establishment of *isomorphism* between the *optimal measurement algorithms* and the *positional numeral systems* (the *binomial codes* and the *Fibonacci p-codes*) [6, 16, 136].

## Chapters 3–7

(7) **General remark**: Chapters 3–7 are devoted to the description of the applications of *algorithmic measurement theory* to computer

science and digital metrology; these chapters include the following scientific results:

## Chapter 3. Evolution of Numeral Systems

(i) Closing the mystery of the *Egyptian calendar* and developing the *dodecahedral–icosahedral doctrine* of the origin of the *Babylonian positional numeral system* with base 60 (the *dodecahedral doctrine*) and the *Mayan numeral system* with base 20 (the *icosahedral doctrine*);

(ii) Justification of *Brousentsov's ternary principle* and the description of the history of the ternary computer "Setun" [72].

## Chapter 4. Bergman's System and "Golden" Number Theory

(i) From the detailed analysis of *Bergman's system*, proposed by the American 12-year-old wunderkind Georg Bergman in 1957 [54], the following important conclusion can be drawn: possibly *Bergman's system* is the most outstanding contemporary mathematical discovery in the field of "*numeral systems*" after the "*Babylonian numeral system*" with base 60 and after the "*decimal*", "*binary*", and "*ternary*" numeral systems. Unfortunately, most mathematicians (including *Bergman* himself) and computer experts of that period (1957) filed to evaluate the revolutionary importance of *Bergman's scientific discovery* for the development of both mathematics and computer science and refer *Bergman's system* to the category of outstanding mathematical results, having a great theoretical or applied importance. The American mathematician Donald Knuth, Professor Emeritus of Stanford University, became the only exception; he posted the link to *Bergman's system* [54] in Knuth's famous book *The Art of Computer Programming* [123], the 20th century scientific bestseller.

(ii) After a detailed analysis of the outstanding 1957 article by Bergman [54], Alexey Stakhov obtained a new fundamental result in the number theory called the "*golden*" *number theory* [94]. Stakhov's conception of the "*golden*" *number theory* led to the establishment of the *new properties of natural numbers* (*Z*- and *D*-properties, *F*- and *L-codes*), following *Bergman's system* [54]. The fact of the establishment of new properties of natural numbers [94], the study of which began 2,000 years ago in ancient Greek mathematics, turned out to be a much unexpected surprise for most mathematicians.

## Chapter 5. The "Golden" Ternary Mirror-Symmetrical Arithmetic

The creation of the *ternary mirror-symmetrical arithmetic*, arising from *Bergman's system* [54], is the most convincing example of the scientific importance of *Bergman's system* for further development of computer science. A conception of the *ternary mirror-symmetrical arithmetic* and its technical realization in the matrix and pipeline mirror-symmetrical arithmetical unit is described in 2002 article by Stakhov [72]. Also, the high appreciation of Stakhov's *ternary mirror-symmetrical arithmetic* by the two outstanding world experts in modern computer science, the American mathematician Donald Knuth, the Professor Emeritus of Stanford University, and the prominent Russian engineer and researcher Nikolay Brusentsov, the Chief designer of the ternary computer "Setun" (Moscow University), should be noted.

## Chapter 6. Fibonacci *p*-Codes and Fibonacci Arithmetic for Mission-Critical Applications

(i) After a detailed analysis of the conception of *"Trojan horse"* of the binary system (its zero redundancy), proposed by the prominent Russian expert in computer science academician Khetagurov, the author Alexey Stakhov came to the conclusion that *humanity is becoming hostage to the binary system for the case of mission-critical applications.*

(ii) To improve the noise immunity of the computers and the metrological stability of the measuring systems for the mission-critical applications, we discuss in Volume II the conception of the redundant *Fibonacci p-codes* and the original *Fibonacci arithmetic*, which are the alternatives to the *binary system* and the *binary arithmetic* and are the basis of the new class of the *noise-immune and fault-tolerant computers*, the *Fibonacci computers for the mission-critical applications* [59–62]. Volume II, considers the Harvard realization of different devices of the *Fibonacci computers*, including the original *Fibonacci counter* for the MINIMAL FORM (the Ukrainian patents "*Borisenko–Stakhov counter*") and the original *Fibonacci arithmetic*, based on the *basic micro-operations* [99, 100].

## Chapter 7. Codes of the Golden $p$-Proportions $(p = 0, 1, 2, 3, \ldots)$

In Chapter 11 of Volume II, a new class of the "*golden*" *redundant numeral systems with irrational bases* (*codes of the golden p-proportions*), which are a *generalization of the binary system* $(p = 0)$ and *Bergman's system* $(p = 1)$, is introduced. The most important scientific results of Chapter 11 are as follows:

(i) The original "*golden*" *arithmetic*, which is a *generalization of the classical binary arithmetic*, is developed. The "*golden*" *arithmetic* retains all the advantages of the *classical binary system* (a simplicity of performing basic arithmetic operations and a simplicity of their technical implementation, a simplicity of the comparison of the "*golden*" *representations* by the value, the possibility of the shift of the "*golden*" *representations* to the left, which to the right, which corresponds to the *multiplication* and *division* of the "*golden*" *representations* in *Bergman's system* [54], finally, a *floating-point representation* of the "*golden*" *representations* in *Bergman's system*) [19].

(ii) It is proved that the *codes of the golden p-proportions* are a source for the original theory of "*golden*" *resistive dividers*, which is the basis of the "*golden*" *digital metrology* and the "*golden*" *self-correcting digital-to-analog and analog-to-digital converters* [58].

# Bibliography

## The Books in the Field of the Golden Section, Fibonacci Numbers and Mathematics of Harmony

[1] Shevelev I.Sh., *Meta-Language of Wildlife*. Moscow: Sunday, (2000) (in Russian).

[2] Shevelev I.Sh., *The Principle of Proportion*. Moscow: Stroiizdat, (1986) (in Russian).

[3] Shevelev I.Sh., Marutaev M.A., Shmelev I.P., *Golden Section. Three Views on the Harmony of Nature*. Moscow: Stroiizdat (1990) (in Russian).

[4] Soroko E.M., *Structural Harmony of Systems*. Minsk: Science and Technology (1984) (in Russian).

[5] Losev A., The History of Philosophy as a School of Thought. *Communist* **11** (1981) (in Russian).

[6] Stakhov A.P., *The Mathematics of Harmony. From Euclid to Contemporary Mathematics and Computer Science*. Assisted by Scott Olsen. World Scientific (2009).

[7] Coxeter H.S.M., *Introduction to Geometry*. New York: John Wiley & Sons (1961).

[8] Vorobyov N.N., *Fibonacci Numbers*. Moscow: Science (1984) (first edition, 1961) (in Russian).

[9] Hoggatt V.E. Jr., *Fibonacci and Lucas Numbers*. Boston, MA: Houghton Mifflin (1969).

[10] Shestakov V.P., *Harmony as an Aesthetic Category*. Moscow: Science (1973) (in Russian).

[11] Vajda S., *Fibonacci & Lucas Numbers, and the Golden Section. Theory and Applications*. Ellis Harwood Limited (1989).

[12] Gardner M., *Mathematics, Magic and Mystery*. New York: Dover Publications (1952).

[13]   Brousseau A., *An Introduction to Fibonacci Discovery*. San Jose, CA: Fibonacci Association (1965).

[14]   Huntley H.E., *The Divine Proportion: A Study in Mathematical Beauty*. Dover Publications (1970).

[15]   Ghyka M., *The Geometry of Art and Life*. Dover Publications (1977).

[16]   Stakhov A.P., *Introduction to Algorithmic Measurement Theory*. Moscow: Soviet Radio (1977) (in Russian).

[17]   Stakhov A.P., *Algorithmic Measurement Theory*. Moscow: Knowledge (1979). (New in Life, Science and Technology. Mathematics and Cybernetics, No. 6) (in Russian).

[18]   Rigny A., *The Trilogy of Mathematics*. Translated from Hungarian. Moscow: World (1980) (in Russian).

[19]   Stakhov A.P., *Codes of the Golden Proportion*. Moscow: Radio and Communications (1984) (in Russian).

[20]   Grzedzielski  J.,  *Energetycno-Geometryczny  kod  Przyrody*. Warszawa: Warszwskie centrum studenckiego ruchu naukowego (1986) (in Polen).

[21]   Garland T.H., *Fascinating Fibonacci: Mystery and Magic in Numbers*. Dale Seymour (1987).

[22]   Kovalev F., *The Golden Ratio in Painting*. Kiev: High School (1989) (in Russian).

[23]   Noise-immune codes. *Fibonacci Computer*. Moscow: Knowledge, Radio Electronics and Telecommunications, No. 6 (1989) (in Russian).

[24]   Vasyutinsky N.A., *Golden Proportion*. Moscow: Young Guard (1990) (in Russian).

[25]   Runion G.E. *The Golden Section*. Dale Seymour (1990) (in Russian).

[26]   Fisher R., *Fibonacci Applications and Strategies for Traders*. New York: John Wiley & Sons, Inc. (1993).

[27]   Shmelev I.P., *The Phenomenon of Ancient Egypt*. Minsk: RITS (1993) (in Russian).

[28]   Bodnar O.Ya., *The Golden Section and Non-Euclidean Geometry in Nature and Art*. Lvov: Sweet (1994) (in Russian).

[29]   Dunlap R.A., *The Golden Ratio and Fibonacci Numbers*. World Scientific (1997).

[30]   Tsvetkov V.D., *Heart, Golden Proportion and Symmetry*. Pushchino (1997) (in Russian).

[31]   Korobko V.I., *The Golden Proportion and the Problems of Harmony of Systems*. Moscow: Publishing House of the Association of Building Universities of the CIS countries (1998) (in Russian).

[32]   Herz-Fischler R., *A Mathematical History of the Golden Number*. New York: Dover Publications, Inc. (1998).

[33] de Spinadel V.W., *From the Golden Mean to Chaos*. Nueva Libreria (1998) (second edition, Nobuko, 2004).

[34] Gazale Midhat J., *Gnomon. From Pharaohs to Fractals*. Princeton, New Jersey: Princeton University Press (1999).

[35] Prechter R.R., *The Wave Principle of Human Social Behaviour and the New Science of Socionomics*. Gainesville, GA: New Classics Library (1999).

[36] Koshy T. *Fibonacci and Lucas Numbers with Applications*. New York: Wiley (2001).

[37] Kappraff J., *Connections. The Geometric Bridge Between Art and Science*. Second Edition. Singapore: World Scientific (2001).

[38] Kappraff J., *Beyond Measure. A Guided Tour through Nature, Myth and Number*. Singapore: World Scientific (2002).

[39] Livio M., *The Golden Ratio: The Story of Phi, the World's Most Astonishing Number*. New York: Broadway Books (2002).

[40] Stakhov A.P., *New Math for Wildlife. Hyperbolic Fibonacci and Lucas Functions*. Vinnitsa: ITI (2003) (in Russian).

[41] Stakhov A.P., *Under the Sign of the "Golden Section". Confession of the Son of Studbat's Soljer*. Vinnitsa: ITI (2003) (in Russian).

[42] Bodnar O.Ya., *The Golden Section and Non-Euclidean Geometry in Science and Art*. Lviv: Ukrainian Technologies (2005) (in Ukrainian).

[43] Petrunenko V.V., *The Golden Section of Quantum States and Its Astronomical and Physical Manifestations*. Minsk: Law and Economics (2005) (in Russian).

[44] Dimitrov V., *A New Kind of Social Science. Study of Self-Organization of Human Dynamics*. Morrisville Lulu Press (2005).

[45] Soroko E.M., *The Golden Section, the Processes of Self-Organization and the Evolution of Systems. Introduction to the General Theory of Systems Harmony*. Moscow: URSS (2006) (in Russian).

[46] Stakhov A., Sluchenkova A., Shcherbakov I., *Da Vinci Code and Fibonacci Numbers*. St. Petersburg: Peter (2006) (in Russian).

[47] Olsen S., *The Golden Section: Nature's Greatest Secret*. New York: Walker Publishing Company (2006).

[48] Petoukhov S.V., *Matrix Genetics, Algebras of Genetic Code, Noise Immunity*. Moscow-Izhevsk: Research Center "Regular and Chaotic Dynamics" (2008) (in Russian).

[49] The Prince of Wales with Tony Juniper and Ian Scelly, *Harmony: A New Way of Looking at Our World*. Harper Publisher (2010).

[50] Arakelian H., *Mathematics and History of the Golden Section*. Moscow: Logos (2014) (in Russian).

[51] Stakhov A., Aranson S., *The Mathematics of Harmony and Hilbert's Fourth Problem. The Way to Harmonic Hyperbolic and Spherical Worlds of Nature.* Lambert Academic Publishing, Germany (2014).

[52] Stakhov A., Aranson S., Assisted by Scott Olsen. *The "Golden" Non-Euclidean Geometry,* World Scientific (2016).

[53] Stakhov A., *Numeral Systems with Irrational Bases for Mission-Critical Applications.* World Scientific (2017).

## The Articles in the Field of the Golden Section, Fibonacci Numbers and Mathematics of Harmony

[54] Bergman G., A number system with an irrational base. *Mathematics Magazine* **31**, (1957).

[55] Stakhov A.P., Redundant binary positional numeral systems. In *Homogenous Digital Computer and Integrated Structures,* No. 2. Taganrog: Publishing House "Taganrog Radio University" (1974) (in Russian).

[56] Stakhov A.P., An use of natural redundancy of the Fibonacci number systems for computer systems control. *Automation and Computer Systems* **6**, (1975) (in Russian).

[57] Stakhov A.P., Principle of measurement asymmetry. *Problems of Information Transmission* **3**, (1976) (in Russian).

[58] Stakhov A.P., Digital metrology in the Fibonacci codes and the golden proportion codes. In *Contemporary Problems of Metrology.* Moscow: Publishing House of Moscow Machine-building Institute (1978) (in Russian).

[59] Stakhov A.P., The golden mean in digital technology. *Automation and Computer Systems* **1**, (1980) (in Russian).

[60] Stakhov A.P., Algorithmic measurement theory and fundamentals of computer arithmetic. *Measurement. Control. Automation* **2**, (1988) (in Russian).

[61] Stakhov A.P., The golden section in the measurement theory. An international *Computers & Mathematics with Applications* **17**(4–6), (1989).

[62] Stakhov A.P., The golden proportion principle: Perspective way of computer progress. *Visnyk Akademii Nauk Ukrainy* **1–2**, (1990) (in Ukrainian).

[63] Stakhov A.P., The golden section and science of system harmony. *Reports of the National Academy of Sciences of Ukraine* **12**, (1991) (in Ukrainian).

[64] Stakhov A.P., Tkachenko I.S., Hyperbolic Fibonacci trigonometry. *Reports of the National Academy of Sciences of Ukraine* **208**(7), (1993) (in Russian).

[65] Stakhov A.P., Algorithmic measurement theory: A general approach to number systems and computer arithmetic. *Control Systems and Computers* 4–5, (1994) (in Russian).

[66] Stakhov A.P., The golden section and modern harmony mathematics. *Applications of Fibonacci Numbers* 7, (1998).

[67] Spears C.P., Bicknell-Johnson M. Asymmetric cell division: Binomial identities for age analysis of mortal vs. immortal trees. *Applications of Fibonacci Numbers* 7, (1998).

[68] Stakhov A.P., Mathematization of harmony and harmonization of mathematics. Moscow: *Academy of Trinitarism*, El. No. 77-6567, publication 166897, (2011) (in Russian).

[69] Stakhov A.P., A generalization of the Fibonacci $Q$-matrix. *Reports of the National Academy of Sciences of Ukraine* 9, (1999).

[70] Stakhov A., Matrix arithmetics based on Fibonacci matrices. *Samara-Moskow: Computer Optics* 21, (2001).

[71] Stakhov A., Ternary mirror-symmetrical arithmetic and its applications to digital signal processing. *Computer Optics* 21, (2001).

[72] Stakhov A.P., Brousentsov's ternary principle, Bergman's number system and ternary mirror-symmetrical arithmetic. *The Computer Journal* 45(2), (2002).

[73] Radyuk M.S., The second golden section (1,465 ...) in Nature. Proceedings of the international conference "Problems of harmony, symmetry and the golden section in nature, science and art. *Vinnitsa State Agrarian University* 15, (2003) (in Russian).

[74] Stakhov A.P., Generalized golden sections and a new approach to the geometric definition of a number. *Ukrainian Mathematical Journal* 56(8), (2004) (in Russian).

[75] Stakhov A., Rozin B., On a new class of hyperbolic function. *Chaos, Solitons & Fractals* 23(2), (2005).

[76] Stakhov A.P., The generalized principle of the golden section and its applications in mathematics, science, and engineering. *Chaos, Solitons & Fractals* 26(2), (2005).

[77] Stakhov A., Rozin B., The golden shofar. *Chaos, Solitons & Fractals* 26(3), (2005).

[78] Stakhov A.P., Golden section, sacred geometry and mathematics of harmony. In *Metaphysics. Century XXI.* Collection of Papers. Moscow: BINOM (2006) (in Russian).

[79] Stakhov A.P., Fundamentals of the new kind of Mathematics based on the golden section. *Chaos, Solitons & Fractals* 27(5), (2006).

[80] Stakhov A., Rozin B., The continuous functions for the Fibonacci and Lucas $p$-numbers. *Chaos, Solitons & Fractals* 28(4), (2006).

[81]    Stakhov A., Fibonacci matrices, a generalization of the "Cassini formula," and a new coding theory. *Chaos, Solitons & Fractals* **30**(1), (2006).

[82]    Stakhov A.P., Gazale formulas, a new class of the hyperbolic Fibonacci and Lucas functions, and the improved method of the "golden" cryptography. Moscow: Academy of Trinitarism, No. 77-6567, publication 14098 (2006) (in Russian).

[83]    Stakhov A., The "golden" matrices and a new kind of cryptography. *Chaos, Solitons & Fractals* **32**(3), (2007).

[84]    Stakhov A.P., The generalized golden proportions, a new theory of real numbers, and ternary mirror-symmetrical arithmetic. *Chaos, Solitons & Fractals* **33**(2), (2007).

[85]    Stakhov A.P., Three "key" problems of mathematics on the stage of its origin and new directions in the development of mathematics, theoretical physics and computer science. Moscow: Academy of Trinitarism, No. 77-6567, publication 14135 (2007) (in Russian).

[86]    Stakhov A.P., The mathematics of harmony: Clarifying the origins and development of mathematics. *Congressus Numerantium* **193**, (2008).

[87]    Stakhov A.P., Aranson S.Ch., "Golden" Fibonacci goniometry, Fibonacci–Lorentz transformations, and Hilbert's fourth problem. *Congressus Numerantium* **193**, (2008).

[88]    Stakhov A.P., Aranson S.Ch., Hyperbolic Fibonacci and Lucas functions, "Golden" Fibonacci goniometry, Bodnar's geometry, and Hilbert's fourth problem. Part I. Hyperbolic Fibonacci and Lucas functions and "Golden" Fibonacci goniometry. *Applied Mathematics* **2**, (2011).

[89]    Stakhov A.P., Aranson S.Ch., Hyperbolic Fibonacci and Lucas functions, "Golden" Fibonacci goniometry, Bodnar's geometry, and Hilbert's fourth problem. Part II. A new geometric theory of phyllotaxis (Bodnar's geometry). *Applied Mathematics* **3**, (2011).

[90]    Stakhov A.P., Aranson S.Ch., Hyperbolic Fibonacci and Lucas functions, "Golden" Fibonacci goniometry, Bodnar's geometry, and Hilbert's fourth problem. Part III. An original solution of Hilbert's fourth problem. *Applied Mathematics* **4**, (2011).

[91]    Stakhov A.P., Hilbert's fourth problem: Searching for harmonic hyperbolic worlds of nature. *Applied Mathematics and Physics* **1**(3), (2013).

[92]    Stakhov A., A History, the main mathematical results and applications for the mathematics of harmony. *Applied Mathematics* **5**, (2014).

[93]   Stakhov A., The mathematics of harmony. Proclus' hypothesis and new view on Euclid's elements and history of mathematics starting since Euclid. *Applied Mathematics* **5**, (2014).

[94]   Stakhov A., The "golden" number theory and new properties of natural numbers. *British Journal of Mathematics & Computer Science* **6**, (2015).

[95]   Stakhov A., Proclus hypothesis. *British Journal of Mathematics & Computer Science* **6**, (2016).

[96]   Stakhov A., Aranson S., Hilbert's fourth problem as a possible candidate on the MILLENNIUM PROBLEM in geometry. *British Journal of Mathematics & Computer Science* **4**, (2016).

[97]   Stakhov A., Fibonacci $p$-codes and codes of the golden $p$-proportions: New informational and arithmetical foundations of computer science and digital metrology for mission-critical applications. *British Journal of Mathematics & Computer Science* **1**, (2016).

[98]   Stakhov A., Aranson S., The fine-structure constant as the physical-mathematical MILLENNIUM PROBLEM. *Physical Science International Journal* **1**, (2016).

[99]   Stakhov A., The importance of the golden number for mathematics and computer science: Exploration of the Bergman's system and the Stakhov's ternary mirror-symmetrical system (numeral systems with irrational bases). *British Journal of Mathematics & Computer Science* **3**, (2016).

[100]  Stakhov A., Mission-critical systems, paradox of hamming code, row hammer effect, 'Trojan Horse' of the binary system and numeral systems with irrational bases. *The Computer Journal* **61**(7), (2018).

## Other Publications

[101]  Kline M., *Mathematics. Loss of Certainty*. Translated from English. Moscow: World (1984) (in Russian).

[102]  Kolmogorov A.N., *Mathematics in Its Historical Development*. Moscow: Science (1991) (in Russian).

[103]  Harmony of spheres, *The Oxford Dictionary of Philosophy*, Oxford University Press (1994, 1996, 2005).

[104]  *The Elements of Euclid. Books I–VI*. Translation from Greek and comments by DD Mordukhay-Boltovsky. Moscow-Leningrad (1948) (in Russian).

[105]  *The Elements of Euclid. Books VII–X*. Translation from Greek and comments by DD Mordukhay-Boltovsky. Moscow-Leningrad (1949) (in Russian).

[106]   *The Elements of Euclid. Books XI–XV.* Translation from Greek and comments by DD Mordukhay-Boltovsky. Moscow-Leningrad (1950) (in Russian).

[107]   Khinchin A.Ya., *Chain Fractions.* Moscow: Fizmatgiz (1961) (first edition, 1935) (in Russian).

[108]   Radoslav J., Pythagoras theorem and Fibonacci numbers http:// milan.milanovic.org/math/english/Pythagoras/Pythagoras.html.

[109]   Korneev A.A., *Structural Secrets of the Golden Series.* Moscow: Academy of Trinitarism, El No. 77-6567, publication 14359 (2007) (in Russian).

[110]   Vilenkin N.V., *Combinatorics.* Moscow: Science (1969) (in Russian).

[111]   Poya D., *Mathematical Discovery.* Translated from English. Moscow: Science (1970) (in Russian).

[112]   Venninger M., *Models of Polyhedra.* Translation from English. Moscow: World (1974) (in Russian).

[113]   Klein F., *Lectures on the Icosahedron and Solving Fifth-Degree Equations.* Moscow: Science (1989) (in Russian).

[114]   Katz E.A., *Art and Science: About Polyhedra in General and the Truncated Icosahedron in Particular.* Moscow: Energy No. 10–12 (2002) (in Russian).

[115]   Gratia D., Quasicrystals. *Uspekhi Fizicheskikh Nauk*, Vol. **156**(2), (1988) (in Russian).

[116]   Eletsky A.V., Smirnov B.M., Fullerenes, *Uspekhi Fizicheskikh Nauk* **163**(2), (1993) (in Russian).

[117]   Kann C.H., *Pythagoras and Pythagoreans. A Brief History.* Hackett Publishing Co, Inc. (2001).

[118]   Vladimirov, Yu.S., *Metaphysics.* Moscow: Binom, Laboratory of Knowledge (2002) (in Russian).

[119]   Vladimirov, Yu.S., Quark icosahedron, charges and Weinberg angle. In *Proceedings of the International Conference "Problems of Harmony, Symmetry and the Golden Section in Nature, Science and Art"*, Vinnitsa (2003) (in Russian).

[120]   Verkhovsky L.I., Platonic solids and elementary particles. *Chemistry and Life* **6**, (2006) (in Russian).

[121]   Zhmud L., *The Origin of the History of Science in Classical Antiquity.* Walter de Gruyter (2006).

[122]   Smorinsky C., *History of Mathematics. A Supplement.* Springer (2008).

[123]   Donald K., *The Art of Computer Programming (TAOCP) in 4th Volumes*, Addison-Wesley (1962, 1968, 1969, 1973, 2005).

[124] Markov A.A., *On the Logic of Constructive Mathematics*. Moscow: Knowledge (1972) (in Russian).

[125] Hilbert D., On the infinite. In *Foundations of Geometry* (1948).

[126] Zenkin A.A., The error of George Cantor. *Problems of Philosophy* **2**, (2000) (in Russian).

[127] Stakhov A.P., Kleshchev D.S., The problem of the infinite in mathematics and philosophy from Aristotle to Zenkin. Moscow: Academy of Trinitarism, El. No. 77-6567, publication 15680, 03.12.2009 (in Russian).

[128] Kleschev D., Pseudoscience: A disease that there is no one to cure. Moscow: Academy of Trinitarismm, El No. 77-6567, publication 17012, 11/22/2011 (in Russian).

[129] Stakhov A.P., Is modern mathematics not standing on the "pseudoscientific" foundation? (Discussion of the article by Denis Kleshchev "Pseudoscience: A disease that there is no one to cure"). Moscow: Academy of Trinitarism, El No. 77-6567, publication 17034, 11/28/2011 (in Russian).

[130] Weyl G., *On the Philosophy of Mathematics*. Moscow-Leningrad, 1934, Reprint Moscow: KomKniga (2005) (in Russian).

[131] Wilde Duglas J., *Optimum Seeking Methods*. Translation from English. Moscow: Science (1967).

[132] Niccolò F.T., From Wikipedia, the free encyclopedia https://en.wikipedia.org/wiki/Niccol%C3%B2_Fontana_Tartaglia.

[133] Bashmakova I.G., Yushkevich A.P., The origin of the numeral systems. In the book "Encyclopaedia of elementary arithmetic". *Arithmetic*. Moscow-Leningrad: State Publishing House of Technical and Theoretical Literature (1951).

[134] Neugebauer O. *Lectures on the History of Ancient Mathematical Sciences. Volume One "Pre-Greek Mathematics"*. Translated from German. Moscow-Leningrad: United Scientific and Technical Publishing House of the USSR (1937) (in Russian).

[135] Pospelov D.A., *Arithmetic Foundations of Computers of Discrete Action*. Moscow: High School (1970) (in Russian).

[136] Borisenko A., *Binomial Counting. Theory and Practice*. Publishing House "University Book", 2004 (in Russian).

[137] Iliev L., Mathematics as a science about models. *Successes of Mathematical Sciences* **27**(2), (1972) (in Russian).

[138] Kartsev M.A., *Arithmetic of Digital Machines*. Moscow: Science (The main edition of the physical and mathematical literature), (1969) (in Russian).

# SERIES ON KNOTS AND EVERYTHING

ISSN: 0219-9769

*Editor-in-charge:* Louis H. Kauffman *(Univ. of Illinois, Chicago)*

The Series on Knots and Everything: is a book series polarized around the theory of knots. Volume 1 in the series is Louis H Kauffman's Knots and Physics.

One purpose of this series is to continue the exploration of many of the themes indicated in Volume 1. These themes reach out beyond knot theory into physics, mathematics, logic, linguistics, philosophy, biology and practical experience. All of these outreaches have relations with knot theory when knot theory is regarded as a pivot or meeting place for apparently separate ideas. Knots act as such a pivotal place. We do not fully understand why this is so. The series represents stages in the exploration of this nexus.

Details of the titles in this series to date give a picture of the enterprise.

More information on this series can also be found at http://www.worldscientific.com/series/skae

Printed in the United States
by Baker & Taylor Publisher Services